Charting Spiritual Care

Simon Peng-Keller • David Neuhold
Editors

Charting Spiritual Care

The Emerging Role of Chaplaincy Records
in Global Health Care

 Springer

Editors
Simon Peng-Keller
University of Zurich
Zurich, Switzerland

David Neuhold
University of Zurich
Zurich, Switzerland

ISBN 978-3-030-47072-2   ISBN 978-3-030-47070-8   (eBook)
https://doi.org/10.1007/978-3-030-47070-8

This Springer imprint is published by the registered company Springer Nature Switzerland AG
The registered company address is: Gewerbestrasse 11, 6330 Cham, Switzerland

Preface

This volume investigates a much debated issue of late-modern spiritual care: the integration of spiritual concerns and chaplaincy in electronic medical records (EMR). This raises several important questions which spiritual care givers have been trying to address from the start. Besides legal and ethical aspects, the ongoing discussion includes the *who, what, where, when, how*, and *why* of recording spiritual care in EMRs.

Until now, there has been a lack of research on this ongoing process and its theological and practical implications. Drawing upon the work of an international study group, this book gives a critical overview of charting practices and experiences in various countries and health care contexts. In particular, we hope to foster greater integration of current developments in spiritual care. Research in this area has been developing steadily over the last three decades. However, the discourses are fragmented, dispersed into different disciplines, and sometimes exist in parallel without any points of contact. There are now countless individual studies, but only a few generally accepted foundations and concepts.

This volume is an attempt to encourage debate on recording spiritual care in EMR and to counteract the dispersion and fragmentation mentioned in three ways. First, by providing an international synopsis: The book encompasses perspectives and models from North America (USA and Canada), Australia, Europe (UK, Netherlands, Belgium, and Switzerland), and Africa (Kenya).

Secondly, we have adopted an interdisciplinary approach, integrating perspectives from theology, psychiatry/psychotherapy, nursing, and bioethics. For the further development of interprofessional records of spiritual care, all these perspectives (and others) are important and have to be brought together in a process of cross-fertilization. That all authors have a Christian background (from different denominations), was not a conscious choice. It reflects the fact that, until recently, discourses on healthcare chaplaincy have mainly been shaped by Christian theologies, although the practice of recording spiritual care in EMR is also spreading in Jewish and Muslim circles.

Thirdly, we put the development into a historical context. By reconstructing the genealogy of a new practice, we aim for a deeper understanding of the current development and its drivers. What the contributions to this volume examine is part of a much wider transformation, which involves the digitalization and globalization of healthcare, the religious-spiritual pluralization of society, as well as the rise of outcome-oriented chaplaincy and interprofessional spiritual care.

Research on a rapidly changing field of practice runs the risk of being quickly overtaken. This may also apply to this volume. It is foreseeable that the field of practice examined here will continue to develop strongly in the coming years. The fact that we consciously take this risk is doubly justified. On the one hand, it is our commitment to contribute to sustainable development in spiritual care through research. On the other hand, a momentary snapshot, such as that provided in the following pages, may be valuable for later research projects.

Zurich, Switzerland   Simon Peng-Keller
David Neuhold

Acknowledgments

The original idea which ultimately led to this volume was not born in an academic context. Rather, it was conceived by members of the Swiss hospital chaplains associations, who recognized the importance of the subject of this book at an early stage and initiated this research project. We would like to thank the organizations for their initiative and for their support in recent years.

This initiative ultimately culminated in an interdisciplinary research project based at the University of Zurich and the Faculty of Theology in Chur and funded by the Swiss National Science Foundation [173202 (cf. http://p3.snf.ch/project-173202)]. This ongoing project examines the current developments in chaplaincy documentation from a practical theological and ethical perspective. In order to provide an overview of international developments, we invited colleagues from different countries and contexts to a workshop in Chur: Brent Peery from the USA, Bruno Bélanger from Canada, Linda Ross and Wilfred McSherry from the UK, Christine Hennequin from Victoria/Australia, Anne Vandenhoeck from Belgium, and Wim Smeets from the Netherlands. The first drafts of the chapters were discussed at the "Charting Spiritual and Pastoral Care" workshop held on 10th and 11th January 2019. The short commentaries that punctuate the volume are a distant echo of the debates that took place at this event, on a snowy winter's day high above the rooftops of the town and in the shadow of the Grison Alps in Switzerland.

We express our profound gratitude to the Faculty of Theology in Chur for its hospitality. We also owe our heartfelt thanks to the Catholic Church in the Canton of Zurich and the Reformed Church of Zurich for their support of the professorship for Spiritual Care at the University of Zurich.

Finally, we are particularly grateful to the Swiss National Science Foundation (SNSF), which has generously funded this project and this book, and to David Dolby and Fabian Winiger for their valuable support during the editing process.

Moreover, we would like to thank Hanspeter Schmitt, Eva-Maria Faber, and Birgit Jeggle-Merz from the Faculty of Theology in Chur; Ralph Kunz from the University of Zurich; and Saara Folini, Claudia Graf, and Livia Wey-Meier. All of them are members of the research group on charting and accompanied, supported, and encouraged us in many ways. Last but not least, thanks are due to Pascal Mösli who did a great job in bringing together all these people interested and engaged in charting spiritual care.

Contents

About the Contributors

Line Beauregard, psychologist and program and research officer at the Centre Spiritualitésanté de la Capitale-Nationale. She completed a master's degree in psychology and a doctorate in social work at University of Laval. Line is involved in CSsanté's research activities and is a member of the editorial board of the journal *Spiritualitésanté*.

Bruno Bélanger, program and research officer at the Centre Spiritualitésanté de la Capitale-Nationale. He completed studies in Theology and Social Sciences at the University of Laval in Quebec City and at the University of Montréal. He is the coordinator of the publication of the journal *Spiritualitésanté* and is responsible for research and training activities at the Centre Spiritualitésanté de la Capitale-Nationale (CSsanté).

Mario Bélanger holds a bachelor's degree in Theology and a master's degree in Spiritual Care from University of Laval. Mario is a professional coordinator for the Centre Spiritualitésanté de la Capitale-Nationale and has been working in spiritual care since 2005. He is a witness to the evolution of spiritual care in recent decades, particularly in the world of outpatient palliative care.

Chantal Bergeron practiced as a head nurse in juvenile psychiatry and long-term care. She holds a bachelor of nursing degree from University of Laval (1978) and a certificate in hospital management from the University of Montréal (1984). She also completed a bachelor's and master's degree in Theology (University of Laval, 1993) and has been working at the Centre Spiritualitésanté de la Capitale-Nationale as a professional coordinator in long-term care and mental health since 2004.

Judy Connolly, MA, MDiv, DMin, has been a chaplain at the University of Minnesota Medical Center since 2000 and a contributor to training programs sponsored by the Center to Advance Palliative Care at Fairview Health Services since 2003. Her 2013 focus group research on "Barriers to Nurse and Paraprofessional

Wellness and Retention in the Quaternary Care Setting" led to her current work supporting Health Fairview staff and providers to enhance resilience and well-being.

Anneke de Vries, Dr., pastor and supervisor in the Radboud University Nijmegen Medical Centre (the Netherlands); supervisor, Bible-drama trainer, lecturer, and tutor at the Protestant Theological University; and guest lecturer at the Faculty of Theology and Religious Studies, Radboud University, Nijmegen.

Eva-Maria Faber, Dr. theol., did her studies in Münster, Freiburg i. Br. and Toulouse. She has been professor of dogmatic and foundational theology at the Faculty of Theology in Chur since 2000, and from 2007 to 2015 she was rector of this Faculty. Her research focuses on ecumenism, theology of the twentieth century, John Calvin, and life and theology.

Eckhard Frick, SJ, is a psychiatrist, psychoanalyst (C.G. Jung-Institute Munich), and Catholic priest. He teaches anthropology at the Munich School of Philosophy (www.hfph.de) and spiritual care at the School of Medicine, Technische Universität München (www.spiritualcare.de). Together with Simon Peng-Keller, he is editor in chief of the journal *Spiritual Care*. Recent co-edited works include *Psychoanalyse in technischer Gesellschaft* and *Spirituelle Erfahrung in philosophischer Perspektive* (De Gruyter).

Paul Galchutt worked for ten years as an inpatient palliative care chaplain. He now serves as the research chaplain at University of Minnesota Health. He was a Transforming Chaplaincy research fellow and is a part of the Interprofessional Spiritual Care Education Curriculum faculty. Along with being a health-care chaplain, he has also been an Evangelical Lutheran Church in America pastor for 23 years.

Christine Hennequin worked as a spiritual care practitioner and a manager in public hospitals. Other roles have included coordinating bereavement support, projects, and consumer participation. She has qualifications in pastoral care, social sciences, frontline management, and project management. In 2009, she joined Spiritual Health Victoria and has managed several projects including the regional and rural development of spiritual care, the development and evaluation of frameworks and guidelines, including guidelines for documenting spiritual care. Her work includes consulting and collaborating with a range of stakeholders to develop and support best practice models of spiritual care within health services.

Guy Jobin, professor of moral theology and ethics and vice dean at the Faculty of Theology and Religious Studies of l'Université Laval (Québec, Canada). He obtained his PhD (2000) in theological ethics from Saint Paul University (Ottawa, Canada). He has held the research chair on Religion, Spirituality and Health at Laval University since 2007. He published in 2012 *From Religions to Spirituality. A Biomedical Appropriation of Religion in Contemporary Hospitals* (2nd edition in

2013) (in French). His current research is on the recent evolution of Spiritual Care in Quebec's healthcare system.

Ralph Kunz worked for some years as a pastor and advisor in the Reformed Church of the Kanton Zurich. Since 2004 he is professor of Practical Theology in the Faculty of Theology at the University of Zurich. His research focuses on liturgy, pastoral care, and spirituality. He is on the board of directors of the Center for Gerontology at the University of Zurich.

Wilfred McSherry, professor in nursing working in a joint appointment between Department of Nursing, School of Health and Social Care, Staffordshire University, and the University Hospitals of North Midlands, NHS Trust, United Kingdom, and part-time professor at VID University College, Bergen, Norway. He has had a career in nursing working as a registered nurse primarily within acute hospital care. He has published extensively in this field with several books and many articles addressing different aspects of the spiritual dimension. He is also a founding and executive member of the British Association for The Study of Spirituality (BASS) and a principal fellow of The Higher Education Academy. In 2012 he was made a fellow of the Royal College of Nursing for his unique contribution to nursing in the areas of spirituality/dignity. He is also leading with Professor Linda Ross on the (EPICC) project.

Pascal Mösli, theologian (MA) and supervisor (MAS). He worked for many years in the hospital pastoral care and initiated the documentation tool Seel:is at the Inselspital, Bern. Today he is responsible for palliative care and special pastoral care at the Reformed Churches Bern-Jura-Solothurn. Pascal is a member of the research group at the chair of spiritual care at the University of Zurich and is self-employed as a lecturer, consultant, and project leader in the field of spirituality and health.

David Neuhold, Dr. theol., member of the research group at the chair of spiritual care at the University of Zurich and a church historian. He is also working as scientific collaborator for the *Swiss Journal of Religious and Cultural History* (www. unifr.ch/szrkg) at the University of Fribourg and teaching there as well as at the University of Lucerne.

Brent Peery, vice president for chaplaincy and spiritual care for the Memorial Hermann Health System. His education includes a bachelor's degree in religion from Baylor University, double master's degrees from Southwestern Baptist Theological Seminary, and doctorate in spiritual formation from Azusa Pacific University. Brent is an active member of the Association of Professional Chaplains; he serves on the Quality Committee and also served as the co-chair of the 2015 *Standards of Practice for Professional Chaplains* task force.

Simon Peng-Keller, Dr. theol., professor of spiritual care at the University of Zurich (www.theologie.uzh.ch/faecher/spiritual-care.html), healthcare chaplain at the palliative care department of the University Hospital of Zurich, and accompanying contemplative retreats. Together with Eckhard Frick, he is editor in chief of the journal *Spiritual Care* and the series *Studies in Spiritual Care*. He has authored numerous publications on Christian spirituality, spiritual care, and spiritual experiences close to death.

Linda Ross, professor of nursing (specializing in spirituality) at the University of South Wales. Her PhD in 1992 was the first to explore nurses' perceptions of spirituality and spiritual care, which she published as a book in 1997. For the last 30 years, she has led numerous research studies on spiritual care in both nursing practice and education, as well as leading training workshops and seminars for healthcare staff internationally. She is a founding member and Secretary for the British Society for the Study of Spirituality and Executive Editor for its affiliated publication, *Journal for the Study of Spirituality*. She is a founding member of the European Spirituality in Nursing Research Network, which is currently leading on a European project titled "Enhancing Nurses Competence in Providing Spiritual Care through Innovation Education and Compassionate Care" (EPICC) (www.epicc-project.eu).

Wim Smeets, Dr., head of the Department of Spiritual Care at the Radboud University Nijmegen Medical Centre (the Netherlands), where he is spiritual caregiver, senior supervisor, and member of the Science Committee of the Professional Union.

Anne Vandenhoeck, assistant professor of pastoral care and diaconal theology in the Faculty of Theology and Religious Studies at the KU Leuven, Belgium. She is chair of the Academic Centre for Practical Theology (KU Leuven) and director of ERICH (European Research Centre for Spiritual Care Givers in HealthCare).

Livia Wey-Meier, theologian (MA) and mediator (University of Fribourg), member of the research group at the Chair of Theological Ethics at the University of Chur/Switzerland.

About the Editors

Simon Peng-Keller is professor of spiritual care at the University of Zurich in Switzerland. He also works as healthcare chaplain at the palliative care unit of the University Hospital Zurich. His current research areas of interest include interprofessional spiritual care and healthcare chaplaincy, visionary experiences near death, spiritual needs in palliative care, and theology of spirituality.

David Neuhold PD Dr., Mag. Theol., is postdoc research associate at the professorship of spiritual care on the Faculty of Theology at the University of Zurich in Switzerland. Dr. Neuhold also is an editor of the *Swiss Journal for Religious and Cultural History* (*SZRKG/RSHRC*) at the University of Fribourg, Faculty of Philosophy, Switzerland.

Introduction

Simon Peng-Keller and David Neuhold

1 Documentation: A Key Issue for Spiritual Care in a Digital Age

At first glance, this book deals with a dry, technical question. The issue of digital records sounds like an unremarkable aspect of our increasingly bureaucratic health-care systems rather than a topic for intriguing discussion. However, our experience of digital records tells a different story. In the academic and pastoral forums in which the issue has been discussed, it has led to intense, foundational, and often emotionally charged debates. Why should such a seemingly mundane subject trigger such lively discussions? At least three reasons can be given: First, documenting spiritual care in electronic medical records (EMR), and the associated training, tools, and collaborative work, may divert chaplains' time and energy away from their primary purposes of personal engagement and spiritual care. One may wonder whether the investment of scant resources in an activity of unclear benefit to patients can be justified. Second, documenting spiritual care touches on the professional identity of healthcare chaplains. For many decades, this has been characterized by a clear demarcation from health professionals, which has manifested itself in nonparticipation in typical health professional practices. Documentation was one of them. While it has long been the professional standard of clinical psychologists and social workers to document their work in medical records, it was until recently an unwritten rule that chaplains should not participate in this task. Pastoral confidentiality was and remains the standard argument for this abstention. In the light of this, one might ask: Doesn't chaplaincy jeopardize its professional identity by now conforming to health professional standards? This point leads to a third issue: the emergence of interprofessional spiritual care. Chaplains are confronted with the fact that nurses,

S. Peng-Keller (✉) · D. Neuhold
University of Zurich, Zurich, Switzerland
e-mail: simon.peng-keller@uzh.ch

© The Author(s) 2020
S. Peng-Keller, D. Neuhold (eds.), *Charting Spiritual Care*,
https://doi.org/10.1007/978-3-030-47070-8_1

1

psycho-oncologists, physicians, and other healthcare providers have started to record information about patients' spiritual needs. The traditional abstention of chaplains from interprofessional documentation is difficult to sustain when health professionals are already recording spiritual matters in their notes. What does it mean for healthcare chaplaincy and its documentation that spiritual care is becoming an interprofessional task?

If discussions about documenting spiritual care in EMR are passionate, this is not only because they raise concerns about chaplains' professional identity and the increasing demands placed on them. There are also positive reasons for this. Many sense in this new practice the opportunity for professionalization and a better integration of the spiritual dimension into healthcare. Among them are the majority of the authors represented in this volume. What unites all of the contributors to this collection, however, is a shared concern for the development of healthcare chaplaincy. The models presented in the following chapters showcase their very different visions for the future of spiritual care. The reader of this volume is invited not only to embark on a journey through different health systems but also to enter a laboratory where future models and tools are discussed and tested. Much is still in development, even though impressive prototypes are already in use.

The urgency of a discussion about the various models and forms of chaplaincy documentation can also be questioned. Are we not giving too much attention to something that is ultimately nothing more than a working aid? However, there is now little doubt that future healthcare chaplaincy will be required to demonstrate the extent to which it benefits patients and that documentation will play a necessary part in this. Given the primacy of patient well-being and the centrality of the personal encounter to the chaplain's vocation, it is essential that the impact of any system of documentation on these be carefully considered. Documentation practices require scientific support and a theoretical foundation. This book is intended as a contribution to this work. The next section sums up the research to date. The main desiderata and some perspectives for future research will be outlined in the last chapter of this book.

2 Research on Documenting Spiritual Care

The research on charting spiritual care in EMRs has been growing in recent years. In the following brief overview, we assign the studies relevant to our topic to five overlapping areas: (a) conceptual questions, (b) spiritual assessment, (c) models of documentation, (d) practices of documentation, and (e) patients' perspectives.

(a) *Conceptual questions:* Robert A. Ruff's 1996 contribution, which has been much quoted since, can be seen as the starting point of the conceptual discussion. The title was programmatic: "Leaving Footprints: The Practice and Benefits of Hospital Chaplains Documenting Pastoral Care Activity in Patients' Medical Records." With their notes, chaplains leave traces of their work. Ruff spoke also of the professional need for the "visibility of chaplains" (Ruff 1996, 390). They carry out their work in an evidence-based environment with an emphasis on accountability. Through their participation in interprofessional record keeping, they demonstrate a more comprehensive view of their work than that of their predecessors, conventionally restricted to religious care in a narrower sense (Cadge et al. 2011). It was in 2007, during the first wave of web-based EMRs, that this "visibility of chaplains" led to a burning debate revolving around the professional identity of healthcare chaplains. It was initiated by the physician Roberta Springer Loewy and her husband, the bioethicist Erich H. Loewy (Springer Loewy and Loewy 2007). In their view, chaplaincy certainly has a place in the clinical environment. Nevertheless, chaplains should be denied the right to access or contribute to patient-centered EMRs. The Loewys criticized the increasing professionalization of healthcare chaplaincy and its ever more systemic integration as problematic. Their argument follows the motto: cobbler, stick to your last, i.e., chaplain, stay in your religious field, and don't invade the medical space! The critical position of the Loewys was addressed by McCurdy (2012) and others. Wintz and Handzo (2013/2015), for their part, draw attention to the main difference between parish clergy and healthcare chaplains. They define "clergy confidentiality" as a referring "to the information that someone seeking forgiveness shares with a clergyperson within the context of ritual confession." As ritual confession only plays a marginal role in the work of healthcare chaplains, the reference to "clergy confidentiality" is misleading. According to Wintz and Handzo, the confidentiality of chaplains should be shaped by their specific duties. They point to the standards of the APC [Association of Professional Chaplains] which state that the passing on of information is allowed if it is "relevant to the patient's medical, psycho-social, and spiritual/religious goals of care." In the same vein, Alex Liégeois speaks of the "application of the relevance filter" (Liégeois 2010, 93). Only information relevant for interprofessional care can be written down.

(b) *Documenting spiritual assessments:* The issues of spiritual assessment and documentation are closely connected. The results of any particular spiritual assessment have to be recorded in appropriate form. This is true as much for chaplains as for healthcare professionals (nurses, McSherry 2008; physicians, Puchalski et al. 2009; social workers, Hodge 2014). Although the literature on spiritual assessment is vast (cf. Balboni 2013; Rumbold 2013), the need for adequate forms and tools for documentation has so far rarely been discussed in this context. An exception is the doctoral thesis of Adams (2015) which investigates the relationship between spiritual assessment and the concrete interventions subsequently carried out. In his qualitative study of the patterns of documentation, Adams concluded that there was a need for a more consistent relationship between the two parameters of assessment and intervention. Hilsman offered an approach based primarily on (his own) experience. He brought into play twenty-two "spiritual needs" to be assessed and pleaded for a narrative, interprofessional record consisting of a few sentences (Hilsman 2017, 157). While Hilsman constructs a coherent tool for assessment and documentation in his own terms, the staff chaplains at The Ottawa Hospital followed the requirements of the College of Registered Psychotherapists of Ontario, which they had recently joined. Here, the development of new templates for electronic charting was embedded in a quality improvement project, which also included a qualitative examination of 104 spiritual care assessments that had been posted on the EMR (Stang 2017). This study is seminal in that it proves that electronic records can be both the object of a study and the instrument for research on healthcare chaplaincy. Because of their model-building approach, both Hilsman and Stang could also be included in the next area.

(c) *Models of documentation:* In developing new models and tools for record keeping, chaplains essentially have two options, narratives or click-boxes. The question touches on the essence of spiritual care. In his paper "Pastoral Products or Pastoral Care? How Marketplace Language Affects Ministry in Veterans Hospitals." Tarris D. Rosell (2006) referred specifically to the terminology used in charting and the power of language to both describe and create realities. But even if a mix of both documentation methods seems advisable and narrative practice is widely seen as the more comprehensive approach, there are voices that point to the merits of using checkboxes (Mösli et al. 2020).[1] Burkhart (2011) discussed the advantages and disadvantages of a Likert-scaled flow sheet with fixed categories compared to progress notes. While the click-box approach saves time and is more discreet, the question arises whether such standardization adequately reflects the self-conception of spiritual caregivers. Do narrative entries not reflect more faithfully the individualized approach of chaplaincy? Rebecca Johnson et al. (2016) advocate the development of new language skills and the cultivation of clarity so that the work of spiritual caregivers can be properly understood and accessible for later reference. The available literature (and the present volume) indicates not only that disparate models for charting spiritual care are in use but also that different models may be most beneficial in different contexts. A strict standardization is still far from being achieved, and may not even be desirable (Tartaglia et al. 2016).

(d) *Practices of documentation:* A fourth group of studies examines current practices of record keeping. The research questions here extend in different directions, reflecting the spectrum from more descriptive to more evaluative approaches. An example of the former is the study of Johnson et al. (2016). The data – more than 400 free-text EMR notes – was taken from a particular and highly specialized place: a 23-bed neuroscience-spine intensive care unit. Through content analysis, recurrent topics are identified. The most common topics were reports of "patient and family practices, beliefs, coping mechanisms, concerns, emotional resources and needs, family and faith support, medical decision making and medical communications" (ibd. 137–139). Chaplains'

---

[1] Cf. the contributions of Peery (Chaplaincy Documentation in a Large U.S. Health System), Vandenhoeck (The Spiritual Caregiver as a Bearer of Stories: A Belgian Exploration of the Best Possible Spiritual Care from the Perspective of Charting), and Smeets and de Vries (Spiritual Care and Electronic Medical Recording in Dutch Hospitals) in this volume.

free-text EMR notes were also studied by Lee et al. (2017). Their goal, though, is not merely descriptive but also evaluative. They question the usefulness of chaplaincy EMR notes for interprofessional communication, for "documentation should provide clinically relevant communication." The result of their study on free-text entries, however, indicates a great need for further development: "The value that chaplains contribute, however, through the depth of their interactions with patients, does not seem to be conveyed in the pattern of clinical documentation we observed" (ibd. 195). Most entries were insufficiently specific or contained only information already available in the EMR. Another 2017 study by Aslakson and others examined how the EMR entries of chaplains address spiritual matters, especially in the environment of intensive care. The study found that spirituality is extremely important to patients in intensive care units and their families but identified various environmental hindrances to the proper completion of spiritual care notes (Aslakson 2017, 653). Aslakson and her colleagues are critical of standardized schemes and checkboxes and prefer free-form notes that are characterized in particular by a certain narrative scope (ibid. 654). They draw on patient assessments and underline the importance of these, which are the topic of the next section.

(e) *Patients' perspectives:* The study of Lee and colleagues also highlights the problem of self-reporting bias. There may be a tendency to overreport perceived positive experiences or events that are more important for the chaplains themselves than for others. In their study of chaplain's reports, Montonye and Calderone observed that they provided more information about the caregivers themselves than about the needs of patients (Montonye and Calderone 2010, 65). The notes reflected the perspective of the chaplains themselves rather than providing information from the patient perspective. In order to correct this self-reporting bias, studies are needed that systematically incorporate the patient's perspective (cf. Snowden and Telfer 2017). A pioneering study along these lines was conducted in French-speaking Switzerland. Tschannen et al. (2014) surveyed 50 patients about their general attitudes toward the interprofessional documentation of chaplains. Significantly, the interviews were led by chaplains. The patients were asked whether they were in principle in favor of the passing on of personal information or whether they considered the idea to be problematic. About 70% of the respondents considered it to be desirable. Patients who were more pessimistic about their health status were more inclined to find the disclosure problematic. Remarkably, the study also found that the patients' attitudes were dependent on the interviewer: the answers varied according to interviewing chaplain.

3   Overview of the Volume

The majority of the studies summarized in the previous section were conducted in the United States, reflecting the fact that the integration of chaplaincy in healthcare is most advanced there. Goldstein and colleagues have captured that deep integration in their survey of the US landscape in 2011. At that time, leading hospitals in the United States had fully integrated spiritual care departments, with chaplains retaining write access to the EMRs in the vast majority of cases (Goldstein et al. 2011). One of the aims of this book is to extend this research to other country contexts. In this book, this expansion of geographical scope begins with a historical exploration. The first chapter traces the history of spiritual documentation from an ancient discovery: the practice of notation as an instrument for spiritual self-care. It ends in an age in which digitalization has already penetrated large parts of the healthcare system and world in which the charting of spiritual care has spread to unexpected places, such as the slums of Nairobi.

This globalization of spiritual care and its documentation are described in more detail in the following chapters. They survey national developments in the United States, Canada, Australia, the United Kingdom, the Netherlands, Belgium, and Switzerland. Earlier versions of these papers were presented and discussed at an international conference in Switzerland. The comments after the contributions pick up the issues raised in these lively discussions.

Let us briefly outline the individual contributions: Brent Peery expounds how healthcare chaplaincy has been charted in the United States in recent years. The focus is on the model which chaplains in the Memorial Hermann Health System in Houston, Texas, work with. The templates used have been constantly revised and have reached a high level of development.

A similar dynamic of development, while more recent, can also be seen in the Canadian model presented in the next chapter. Bruno Bélanger and colleagues trace the careful process of reflection, construction, and implementation over the last few years. More than the other models considered in this book, the Canadian charting tool has been constructed around a theological core.

In the following paper, the perspective changes to a more administrative view on the development. As in Canada, new structures for healthcare chaplaincy are also emerging in Australia. Typical for a period of transition, divergent models and practices are being employed side by side. Remarkably, the officials of the administrative body "Spiritual Health Victoria" have fostered and steered this change by issuing guidelines as well as by implementing the "pastoral care intervention codes" in the Australian version of the ICD-10.

Finally, we turn to Europe, starting with a contribution from the United Kingdom which introduces also another professional perspective. Wilfred McSherry and Linda Ross, both rooted in nursing studies, offer an assessment tool for spiritual care in the field of general care. The fraught standing of religion(s) in the United Kingdom complicates the integration of chaplains into an interprofessional spiritual care. Standardized procedures and tools are still lacking. McSherry and Ross point out the importance of appropriate language and terminology.

The Dutch context is represented by Wim Smeets and Anneke de Vries. While interprofessional spiritual care and the integration of chaplains are more advanced here than in the United Kingdom, it is still a contentious whether chaplains' records should be part of the EMRs. The authors see this development as a facet of the professionalization of spiritual care which ultimately serves the well-being of patients.

With Anne Vandenhoeck's paper on the situation in Flanders/Belgium, the circle starts to close. The background of the model introduced at the University Hospital of Leuven accords with that of the Memorial Hermann Hospital in Houston. It builds on the outcome-oriented chaplaincy developed in the 1990s by Arthur Lucas in St. Louis/Missouri. Highlighting the importance of language and terminology, Vandenhoeck advocates a "narrative approach."

Last but not the least, Pascal Mösli offers a view on a small but nonetheless highly diverse context: the microcosmos of healthcare chaplaincy in Switzerland. On the basis of a survey, the contribution summarizes the viewpoints of chaplains. The majority see the current developments as an opportunity for professionalization. Finally, Mösli gives us an insight into a "construction site" where much is in motion and still to be decided.

The contributions in the second half of the book take up key topics already alluded to in the first part. These include the relationship between pastoral and psychological-psychiatric documentation and the spiritual aspects of the latter. Psychiatrists are used to keeping medical records, but not on spiritual issues. With a new openness for religious-spiritual aspects in psychiatry and psychotherapy, the question of the documentation of spiritual aspects arises. Eckhard Frick pleads in his chapter for a hermeneutic approach and speaks of a "translation work" of the psychiatric and psychotherapeutic guild, which has to be done in the documentation.

Most of the documentation models presented in this volume are cross-sectoral. At the same time, it is undisputed that documentation practices should be well adapted to the specific needs of a particular care area or profession. Paul Galchutt and Judy Connolly's contribution is dedicated to the field of palliative care, which plays a key role in the development of interprofessional spiritual care. What expectations do team members have of chaplains' record keeping? The paper makes it clear that chaplains must navigate a broad set of demands.

Finally, Guy Jobin approaches the issue from an ethical perspective. He addresses, among other things, the challenge of standardized information and documentation and the fears that the patient might disappear within the clinical setting. In principle, the "improvement of a deepened clinical relationship" should be the benchmark of technologization, digitalization, and, last but not the least, documentation.

In his final synopsis, Simon Peng-Keller concentrates on the analysis of the main drivers of the development and on perspectives for the future. He concludes that there have been three main drivers which have led to the rapid development of charting spiritual care in EMRs: first, the rise of outcome-oriented chaplaincy; second, the digitalization of society and healthcare; and, third, the religious-spiritual pluralization of Western societies and the subsequent "new governance in religious affairs."

References

Adams, Kevin Eugene. 2015. Patterns in chaplain documentation of assessments and intervention. In *A descriptive Study*. Richmond: VCU Scholars Compass.

Aslakson, Rebecca, Josephine Kweku, Malonnie Kinnison, Sarabdeep Singh, Thomas Y. Crowe, and the AAHPM Writing Group. 2017. Operationalizing the measuring what matters spirituality quality metric in a population of hospitalized, critically ill patients and their family members. *Journal of Pain and Symptom Management* 53 (3): 650–655.

Balboni, Michael J. 2013. A theological assessment of spiritual assessments. *Christian Bioethics* 19 (3): 313–331.

Burkhart, Lisa. 2011. Documenting the story: Communication within a healthcare team. *Vision* 21 (3): 28–33.

Cadge, Wendy, Katherine Calle, and Jennifer Dillinger. 2011. What do chaplains contribute to large academic hospitals? The perspectives of pediatric physicians and chaplains. *Journal of Religion and Health* 50: 300–312.

Goldstein, H. Rafael, Deborah Marin, and Mari Umpierre. 2011. Chaplains and access to medical records. *Journal of Health Care Chaplaincy* 17 (3–4): 162–168.

Hilsman, Gordon. 2017. *Spiritual care in common terms: How chaplains can effectively describe the spiritual needs of patients in medical records*. London/Philadelphia: Jessica Kingsley Publishers.

Hodge, David. 2014. *Spiritual assessment in social work and mental health practice*. New York: Columbia University Press.

Johnson, Rebecca, M. Jeanne Wirpsa, Lara Boyken, Matthew Sakumoto, George Handzo, Abel Kho, and Linda Emanuel. 2016. Communicating chaplains' care: Narrative documentation in a neuroscience-spine intensive care unit. *Journal of Health Care Chaplaincy* 22 (4): 133–150.

Lee, Brittany M., Farr A. Curlin, and Philip J. Choi. 2017. Documenting presence: A descriptive study of chaplain notes in the intensive care unit. *Palliative & Supportive Care* 15: 190–196.

Liégeois, Alex. 2010. Le conseiller spirituel et le partage d'informations en soins de santé. Un plaidoyer pour un secret professionnel partagé. *Counseling et Spiritualité* 29 (2): 85–97.

Loewy, Roberta Springer, and Erich H. Loewy. 2007. Healthcare and the hospital chaplain. *Medscape General Medicine* 9/1. (e-Journal).

McCurdy, David B. 2012. Chaplains, confidentiality and the chart. *Chaplaincy Today* 28/2. (e-Journal).

McSherry, Wilfred. 2008. *Making sense of spirituality in nursing and health care practice, an interactive approach*. London: Jessica Kingsley Publishers.

Montonye, Martin, and Steve Calderone. 2010. Pastoral interventions and the influence of self-reporting: A preliminary analysis. *Journal of Health Care Chaplaincy* 16: 65–73.

Mösli, Pascal, Livia Wey-Meier, and David Neuhold. 2020. Ankreuzen oder erzählen? Ein Pladoyer fur Checkboxen in der klinischen Seelsorgedokumentation. In *Dokumentation als seelsorgliche Aufgabe. Elektronische Patientendossiers im Kontext von Spiritual Care*, ed. Simon Peng-Keller, David Neuhold, Hanspeter Schmitt and Ralph Kunz. Zurich: TVZ (in press).

Puchalski, Christina, et al. 2009. Improving the quality of spiritual care as a dimension of palliative care: The report of the consensus conference. *Journal of Palliative Medicine* 12 (10): 885–904.

Rosell, Tarris D. 2006. Pastoral products or pastoral care? How marketplace language affects ministry in veterans hospitals. *The Journal of Pastoral Care & Counseling* 60 (4): 363–367.

Ruff, Robert A. 1996. "Leaving footprints." The practice and benefits of hospital chaplains documenting pastoral care activities in patients' medical report. *Journal of Pastoral Care* 50/4: 383–391.

Rumbold, Bruce. 2013. Spiritual assessment and health care chaplaincy. *Christian Bioethics* 19 (3): 251–269.

Snowden, Austyn, and Iain Telfer. 2017. Patient reported outcome measure of spiritual care as delivered by chaplains. *Journal of Health Care Chaplaincy* 23 (4): 131–155.

Stang, Vivian B. 2017. An e-chart review of chaplains' interventions and outcomes: A quality improvement and documentation project. *The Journal of Pastoral Care & Counseling* 7/3: 183–191.

Tartaglia, Alexander, Diane Dodd-McCue, Timothy Ford, Charles Demm, and Alma Hassell. 2016. Chaplain documentation and the electronic medical record: A survey of ACPE residency programs. *Journal of Health Care Chaplaincy* 22 (2): 41–53.

Tschannen, Olivier, Pierre Chenuz, Emmanuel Maire, Daniel Petremand, Peter Vollenweider, and Cosette Odier. 2014. Transmission d'informations par les aumôniers dans le dossier-patient: le choix des patients. *Forum Medical Suisse* 14 (49): 924–926.

Wintz, Sue, and George Handzo. 2015. Dokumentation und Verschwiegenheit in der professionellen Seelsorge. *Wege zum Menschen* 67/2: 160–164. See also an English version in: http://www.handzoconsulting.com/blog/2013/9/19/documentation-and-confidentiality-for-chaplains.html.

A Short History of Documenting Spiritual Care

Simon Peng-Keller and David Neuhold

On January 5, 1972, Frère Roger, founder and prior of the Taizé Community, noted down: "I want to pick up my diary again. There is no substitute for writing, this slow reflection, in which I record my efforts and insights in rounded letters on the page that shines in the light of the lamp" (Schutz 1974, 116). Writing notes on his experiences as a spiritual companion and leader served as an instrument of spiritual self-care. In his daily notes, Frère Roger reflected upon experiences, encounters, and decisions. Nonetheless, his writing was in the service of memory – his own as well as that of others. Recording his "efforts and knowledge" was a form of witnessing and pastoral care. Written "in rounded characters on the page that shines in the light of the lamp," Frère Roger's notes are situated in pre-digital modernity. The spiritual caregivers on whom this book focuses and for whom it is written type their notes on keyboards which transmit them to illuminated screens. Their medium, context, and addressees differ from those of Frère Roger. But in one central respect, they take up his experience: in the written retracing of spiritual accompaniment.

Although Frère Roger's documentation practice belongs to the modern age, it resembles forms that have been cultivated in the Christian tradition for centuries. In the following pages, the long history of spiritual documentation, from its very beginnings to the present day, which lives on in recent history, will be brought to mind. Referring to ancient and early modern practices as well as on the developments in the twentieth and twenty-first centuries, we outline a genealogy of charting spiritual care. Historical knowledge affords the possibility of distancing oneself from current ambivalences and quandaries. And sometimes it opens up new perspectives for future development.

S. Peng-Keller (✉)
Spiritual Care, University of Zurich, Zurich, Switzerland
e-mail: simon.peng-keller@uzh.ch

D. Neuhold
University of Zurich, Zurich, Switzerland

© The Author(s) 2020
S. Peng-Keller, D. Neuhold (eds.), *Charting Spiritual Care*,
https://doi.org/10.1007/978-3-030-47070-8_2

11

Historically, literacy has been an agent of change. As Egyptologist Jan Assmann puts it: "Writing is a technology that makes cultural creations possible that would otherwise never exist, and that preserves cultural creations in memory, making accessible to later recourse what would otherwise be forgotten and have vanished" (Assmann 2012, 380). Without *writing*, there are no postaxial "world" religions – and no healthcare chaplaincy. From the very beginning, the extension of communication through external (i.e., written) storage has had a flip side as well: "As with all the more complex instruments, writing [...] gives rise to a dialectic of expansion and loss," Assmann states; since "as an externalized memory, it facilitates a hitherto undreamed-of expansion in our capacity to store and retrieve information and other forms of communication, while simultaneously leading to a shrinkage of our natural memory bank" (Assmann 2011, 9).

Healthcare chaplains struggle with their own version of this dialectic, which is nothing new for religions with a great affinity to writing and reading. In the evolution of Christian spiritual care, there has been, again and again, a strong desire to make systematic records. Writing raises the meaning of life experiences to a new hermeneutic level of reflection. Whereas Frère Roger or Georges Bernanos' diary-writing country priest was emblematic of the twentieth century, healthcare chaplains' writing entries into the ward office computer might play a similar role for spiritual care in the early twenty-first century.

Taking notes in the service of spiritual (self-) care has a long and complex history. As the quotation from Frère Roger's diary shows and as we will outline in more detail, this practice serves at least three purposes: First, it fosters awareness and reflexivity; second, it supports the memory; and, third, it facilitates communication. Our genealogical approach is divided into four sections:

(1) Historical antecedents: Note-writing in Christian spirituality
(2) Note-writing in Clinical Pastoral Education of the twentieth century
(3) Charting spiritual care in electronic medical records (EMRs)
(4) Worldwide distribution

1   Historical Antecedents: Writing Notes and Christian Spirituality

In order to shorten the long prehistory of the practices studied in this volume, we concentrate on two typical forms of note-taking in Christian spirituality. While the first one, mentioned in Athanasius' *Vita Sancti Antonii* (Athanasius, ed. Deferrari 1981) about 360 A. D., was intended for spiritual *self*-care, the second example, to be found in the "Roman" structured Jesuit order (Friedrich 2007), focused on the spiritual guidance of a large community. In both cases, writing was connected with remembrance (of past events), reflection (about events in the past and present), and recording (for the future). While note-taking in the first case was a form of accounting for oneself, in the second example, it was more concerned with the administration of a collective. In these two cases, we can see the poles of a broad spectrum: at one pole, notes are taken for the purpose of self-examination and spiritual growth. At the other, note-taking serves institutional purposes.

The spiritual practice of recording, to which Athanasius bears witness, was in all probability inspired by Stoic philosophers who recommended written self-reflection. According to the analyses of Michael Foucault, this practice, refined in the first and second centuries A.D., had an "ethopoietic" function (Foucault 1997, 207–222). In collecting the mind, it fosters meditative self-awareness and transforms the insights gained into practical knowledge. Christian ascetics and hermits like Antonius continued this practice in their own way. The pioneer of all desert fathers, Athanasius reports, invited his followers: "Let us note and write down our deeds and the movements of our soul as if we were to tell them to each other" (Athanasius, Chapter 55, ed. Deferrari 1981, 185). Self-perception through writing presents itself as a spiritual therapeutic process. The written text becomes a counterpart and mirror. Writing serves the ascetic goal: clearing the mind from distracting thoughts and opening it for contemplation.

More than 1000 years later, the former soldier Ignatius of Loyola also attached central importance to spiritual writing. His spiritual life started with an eremitic period and with experiences similar to those of Antonius. The ideas he later systematized in his *Spiritual Exercises* had their origins in his spiritual self-care which included written self-reflections. Later in his career, Ignatius became the "superior general" of an expanding order. Like a Calvin in Geneva or a Bullinger in Zurich, he established an extensive bureaucracy and, importantly for us, a refined system of documentation. His successors continued this model, in which systematic recording and the steady flow of information played an important role. Over time, an information management system was established which encompassed comprehensive and centralist standards. This can be seen, for example, in the personnel catalogues of the Society of Jesus itself, where "evaluation templates" for the identification of different characteristics and qualities of members of the order were common. This was implemented by means of short descriptors ("bene," "valde," "optime," etc.) not dissimilar to those used in current "gap texts" or click boxes. Markus Friedrich states in this context: "The assessment of individuals was thus based on a standardized scale" (Friedrich 2007, 69).

The spiritual guidance administered by sophisticated documentation procedures is more reminiscent of what Foucault described as pastoral power than of an individualized spiritual care. To put it mildly: "[...] the border between 'administrative' and 'edifying' (or: religious) communication was often blurred" (Friedrich 2007, 69). Weighing the different aspects of this documentation practice against each other, Friedrich concludes: "And, of course, for the Jesuits, the efficient organization and administration of their own social body was, in the end, a deeply religious task. They thought about their social body in secular (i.e. administrative) terms, but all for the sake of a religious goal. It might be the ability to combine both perspectives that made the Society of Jesus a successful global player" (Friedrich 2007, 72–73). The success of the Jesuit order can be explained by the combination of different factors. Documentation with ink is one of them.

While Antonius advocated spiritual self-care through writing in an eremitic desert context, Ignatius and his successors were engaged in the successful administration of a global enterprise for which written documentation and communication were crucial. However, writing for self-care and recording for and about others shouldn't be seen as opposing phenomena. As Frère Roger reminds us, the same practice can serve distinctive ends.

2   Note-Writing in Clinical Pastoral Education of the Twentieth Century

With the rise of Clinical Pastoral Education, a new chapter opens up in the history of note-taking for pastoral purposes. This new development took shape in the first decade of the twentieth century in the context of the Emmanuel Movement. It was initiated in Boston by the theologian Elwood Worcester and was influenced by the psychology of William James and Wilhelm Wundt as well as by the New Thought (Hart and Div 2010, 541–542). This Christian pioneer movement, active during the years 1906–1929, started with the aim of providing social and medical support to patients with tuberculosis. Incrementally, the focus turned to nervous disorders and alcoholism, and the spectrum of therapeutical approaches used likewise broadened. Medical instruction as well as psychological and psychotherapeutic approaches (e.g., suggestion, hypnosis) were combined with spiritual support (e.g., confession, prayer, counseling). There was an outpatient service for individual and group therapy connected to the Emmanuel Church, a Presbyterian parish church – hence the name. Interprofessional cooperation was sought and promoted.

To ensure good therapy, it was important for Worcester and his medical colleagues that "pain of all kinds" were diagnosed in the course of treatment, and the "preservation of records, without which no treatment can be regarded as scientific or even safe," was seen as constitutive. The records written by pastors or chaplains were considered essential. For the sake of those in therapy with the Emmanuel Movement, the documentation system of the Massachusetts General Hospital was adopted – "supplemented by notes on the moral and spiritual advice given and on the effect of this advice" (Worcester et al. 1908, 6). Since its foundation in 1821, the Massachusetts General Hospital had played a pioneering role in clinical documentation in general (Gillum 2013, 854). According to Worcester and his colleagues, the collection of information on patients – including spiritual course notes – explicitly served to track the ongoing therapy of people with "nervous disorders" as successfully as possible. The reuse of records was therefore implicitly taken into consideration and regarded as indispensable in and for the therapeutic process.

Although the Emmanuel Movement was a temporary experiment, it inspired further initiatives, especially the Clinical Pastoral Education movement (CPE). The CPE, which started in 1925 in Boston, was linked to that movement, both ideologically and personally. The hematologist Richard Cabot, co-founder of the CPE, was one of the doctors who supported the Emmanuel Movement from the beginning (Hendrick 1914, 410–417). According to Cabot, the spiritual dimension should be included in interprofessional care: clinically trained chaplains should be part of the treatment team and document their experiences in medical records. Right from the beginning, the CPE encompassed forms of note-taking for at least three different purposes: for personal reflection, supervision, and interprofessional communication. While the pioneers of the CPE agreed that written documentation was an essential part of a chaplain's professional role, they had different ideas about how this should be done. Remarkably, these differences mirrored distinctive visions concerning the main task of healthcare chaplaincy. From his medical perspective, Cabot saw the duty of chaplains as consisting in the provision of religious and moral support. Anton Boisen's vision of chaplaincy was more ambitious: he was convinced that chaplains also have a scientific and therapeutic task. He therefore emphasized the hermeneutical function of "careful recordkeeping as means to a 'more conscious and intelligent [religion] capable of verification and transmission'" (Myers-Shirk 2009, 33). In order to understand the "living human documents" entrusted to their care, chaplains need the medium of written documents.

An elaborate description of the different applications of note-taking is to be found in the volume *The Art of Ministering To The Sick*, which Cabot published in 1936 together with the pastor Russell L. Dicks (1906–1965). This publication aimed to close a gap at the interface of religion and medicine by opening up new interdisciplinary horizons. Both authors attribute a decisive role to the written documentation of "ministry," i.e., chaplaincy (cf. Cabot and Dicks 1936, especially chapter "Note-Writing" in part IV "Methods," 244–261). In conclusion, both stated that anyone not writing notes to accompany his work was not suitable for an assignment as hospital chaplain. Expressing it the other way round, in a somewhat gentler formulation, Cabot and Dicks state: "It is inconceivable to us that any conscientious minister can omit writing notes in some form or other" (Cabot and Dicks 1936, 269). Without note-taking, no professional chaplaincy.

Cabot and Dicks focused strongly on the creative power of writing. As in antique self-care, writing notes would have a spiritual function. It can be seen a "spiritual exercise" (Cabot and Dicks 1936, 261). Documentation enables a chaplain to develop further his own work and even to capture something of the transcendent. "Note-writing finds holes and plans to fill them up. It is self-criticism. It is self-revelation. It is preparation for self-improvement" (Cabot and Dicks 1936, 248). In an environment where chaplains are constantly interacting with doctors, such records provide the badge of professionalism. In their book, both authors are more interested in writing for personal usage. Although interprofessional aspects are touched upon[1] and a strong patient orientation is presented, the chaplain himself is in the central field of vision. "Note-writing" serves not only to monitor one's own

---

[1] The interprofessional side will then be more strongly integrated within Russell Dicks' "Standards for the Work of the Chaplain in the General Hospital" (1940), especially in point 4. Text taken from www.professionalchaplains.org/files/professional_standards/standards_of_practice/standards_for_work_of_chaplain_russell_dicks.pdf (without page numbers).

work. It is first and foremost a creative process. It opens up new issues and leads to important insights.

In 1940, four years after *The Art of Ministering To The Sick*, Dicks published his "Standards for the Work of the Chaplain in the General Hospital" (Dicks 1940). This programmatic paper includes a paragraph on recording. Dicks distinguishes here between three main forms[2], daily records "which the chaplain keeps for his own use as a check against his memory"; more detailed records "used especially in difficult assignments to help objectify the patient's need in the chaplain's mind and to show him the mistakes and failures he has made in his work"; and finally notes "in the clinical or medical record itself" for interprofessional communication:

> This is a brief note, similar to that which the consultant writes, which is simply a record of the chaplain's impression of the patient. The chaplain often discovers significant things about a patient which the physician needs to know; these discoveries as well as impressions should be available in the record. Such a note does not reveal confidences which may have been shared with the chaplain nor does it, in any way, infringe upon the sacred nature of the confessional (Dicks 1940, [4]).

It is difficult to estimate the extent to which such practices spread in the following decades. Dicks remarks that this practice "is not generally accepted" and the sparsity of documents hints to a rather slow development. The international reception of the CPE didn't extend to its emphasis on note-taking and record keeping. In Switzerland and Germany, as far as we know, entries concerning pastoral care were restricted to short entries concerning the last rites for Catholic patients.

A new era of clinical pastoral documentation began with the emergence of outcome-oriented chaplaincy in the 1990s. It was a new paradigm of healthcare chaplaincy, one which responded to changes in healthcare generally (e.g., the introduction of the DRG [diagnosis-related group] Codes). Key to this new development was the Barnes-Jewish Hospital in St. Louis, the largest hospital in the USA state of Missouri and teaching hospital for Washington University School of Medicine. "Since 1990," Methodist chaplain and pioneer Arthur M. Lucas summarizes: "the Chaplains at Barnes-Jewish Hospital (BJH) have sought to increase their integration into and accountability with the care teams" (VandeCreek and Lucas 2001, 2). Communicating more effectively with others about spiritual care has been a cornerstone of this new approach, which entails entering the chaplain's assessments, plans, interventions, and evaluations into the medical records. The influence of this initiative is also documented in the present volume. The models presented by Brent Peery (Memorial Hermann Hospital, Houston) and by Anne Vandenhoeck (University Hospital, Leuven) have been shaped by the approach Lucas and his colleagues developed in St. Louis.

---

[2] Dicks mentions also "periodic reports, preferably written reports, to the hospital administrator, to the board of directors of the hospital, and to the church authorities under whose auspices he is serving", Dicks 1940, [5].

3   Charting Spiritual Care in Electronic Medical Records

The emergence of the digital age is the central turning point for the genealogy we are about to trace here. The developments outlined so far achieved a new scope with the introduction of the electronic medical record (Gillum 2013). Let us briefly summarize the history of this cornerstone of digital healthcare.

Efforts to create electronic medical records (EMRs) began in the 1960s, with the promise of transferable documentation that would allow a third party to understand a diagnosis based on detailed entries, from family history to patients habits and blood analysis (Doyle-Lindrud 2015). The 1970s saw the electronic implementation of the so-called problem-oriented medical record (POMR). This system of records, designed by doctor and researcher Larry Weed (1923–2017), made the medical history of a person, his/her so-called problem list, accessible to multiple physicians. Nevertheless, it was only with the rising importance of personal computers that the breakthrough came in the 1990s. A resulting increase in transparency, portability, and accessibility to personal medical data was seen first in hospitals and then in smaller medical environments, such as doctors' surgeries. Data was then made more readily accessible to multiple actors, with the consequence that multiple actors participate in the file in a user-friendly and easy way. An increase in transparency, portability, and accessibility to personalized medical data followed up. The rise of the Internet set new standards and created new possibilities, which then led to web-based EMRs. Around 2010, the Obama administration began to push the development of EMRs in the USA, with repercussions all over the world.

While digitization was initially characterized by the simple replacement of the pen with the keyboard, web-based EMRs bring with them significant qualitative change. The intention was to collect and merge data in a patient-centered manner. Furthermore, clinical data was to be immediately available. A "location-independent" virtual application for different health processes was thus made possible.

Having first appeared in experimental projects at the turn of the millennium, web-based EMRs became more sophisticated and reliable in the 2010s. This was also when governmental healthcare organizations, hospitals, and health insurance companies began to promote and fund the development and implementation of EMRs. The further development of EMRs remains a central challenge today, for example, in Switzerland, following the 2015 legislation to implement personal electronic dossiers from 2020. Questions of security, patient contributions, big data applications, data storage and synchronization, as well as practicability play a major role in this process. Concerning EMRs "physicians also noted dissatisfaction with poor usability, time-consuming data entry, interferences with face-to-face patient care, inefficient and less fulfilling work content, inability to exchange health information" (Doyle-Lindrud 2015, 154). EMRs are also sometimes heavily criticized as instruments which make patients disappear (Hunt et al. 2017).

4   Worldwide Distribution of Spiritual Care Records in EMR

As this volume shows, the worldwide integration of spiritual topics into the EMR has proceeded at an astonishing pace in recent years. To the examples collected in the contributions to this volume, we should like to add another. It comes, rather surprisingly, from a slum in Nairobi.[3] In 2002 the Eastern Deanery AIDS Relief Program (EDARP) initiated electronic record keeping to meet the reporting requirements of this rapidly expanding program. In 2018, Richard W. Bauer, a Maryknoll priest, social worker, and board-certified chaplain, took office in this program with the goal of creating "a *comprehensive* public health intervention that includes assessment and treatment for each individual person's physical, emotional, social, and spiritual well-being and health. Like all health interventions, for quality patient care these interventions (in all domains) must be documented and shared with an interdisciplinary team." There was unanimous assent that EDARP needed to incorporate this into the system of comprehensive, differentiated patient-centered care for HIV and that this assessment needed to be documented in the EMR. It was decided, that the EDARP social workers (one at each of 14 sites) would be the staff initially trained in this methodology and intervention/screening.

> Working with our biostatistician, we concluded that we needed both qualitative and quantitative data. We incorporated the FICA[4] tool and some other quantitative measures to help the social worker and interdisciplinary team better understand how a patient's individual spiritual beliefs and practices may enhance, or impede, adherence to treatment for HIV. The goal of these conversations is not for the social worker to impose their beliefs on the client, but to establish that EDARP values discussing their faith, beliefs and spiritual practices and how these beliefs may impact the individual's treatment and health. Ultimately, we hope to present (but never impose) a model of care and treatment that is about 'prayer and medicine', not, 'prayer or medicine'.

The social workers all began to implement this intervention in the second half of 2018. Richard Bauer meets them frequently to give them support and supervision and to discuss the use of this tool and their impressions regarding acceptability. Overwhelmingly, with few exceptions, patients appreciated the fact that the issue had been brought up.

> I was particularly concerned with one of our sites that is in a predominantly Muslim area of Nairobi and most of our staff identify as Christian. However, the social worker in this site frequently contacted me asking how to cut off the conversations, as patients were so excited to be able to talk with someone about their faith and health that the social worker needed to contain the long conversations.

On January 1, 2019, after a couple of revisions, the tool was then uploaded into the EDARP electronic medical record. Only three of the fourteen sites were to use this version in the EMR, while the other sites continued to use the tool with written

---

[3] We thank Richard W. Bauer for all the information about this project and the opportunity to use them for our historical sketch. The quotations are taken from his personal account, with permission.
[4] Cf. Borneman et al. (2010).

records, sharing information with other clinicians during the interdisciplinary team meetings. The main benefit of having the spiritual screening in the EMR is that all members of the clinical team (nurses, clinical officers, pharmacists, adherence counselors, and social workers) have access to the information. They don't have to repeat screenings and assessments, and they can draw on previous records, obtained from the social worker, in later clinical discussions.

It may be incidental that our historical sketch started in Egypt and ended in Kenya. We came across Bauer's initiative by accident in the context of a research project investigating the WHO's handling of the "spiritual dimension" of health, a discourse in which delegates from the southern hemisphere have played a crucial role. We regard this accidental discovery as a reminder that any future history of charting spiritual care would also have to include developments in the global south. The fact that the use of EMRs with entries on spiritual care has reached the slums of Nairobi (and most probably also other places in the "majority world") is a clear sign of the globalized character of today's healthcare. In conjunction with the ongoing digitalization of social life, the globalization of healthcare will most probably be a main driver for the future development and dissemination in this area.

References

Assmann, Jan. 2011. *Cultural memory and early civilization. Writing, remembrance, and political imagination.* Cambridge, MA: Cambridge University Press.
———. 2012. Cultural memory and the myth of the axial age. In *The axial age and its consequences*, ed. Robert N. Bellah and Hans Joas, 365–407. Cambridge, MA/London: Belknap Press of Harvard University Press.
Athanasius. 1981. The life of St. Anthony. In *Early Christian biographies*, ed. Roy J. Deferrari, 3rd ed., 133–216. Washington, DC: The Catholic University of America Press.
Borneman T. B. Ferrell, and C. M. Puchalski. 2010. Evaluation of the FICA Tool for Spiritual Assessment. *Journal of Pain and Symptom Management* 40/2:163–173. https://doi.org/10.1016/j.jpainsymman.2009.12.019
Cabot, Richard C., and Russell L. Dicks. 1944 [1936]. *The art of ministering to the sick.* New York: The Macmillan Company.
Dicks, Russell L. 1940. *Standards for the work of the Chaplain in the general hospital.* http://www.professionalchaplains.org/files/professional_standards/standards_of_practice/standards_for_work_of_chaplain_russell_dicks.pdf.
Doyle-Lindrud, Susan. 2015. The evolution of the electronic health record. *Clinical Journal of Oncology Nursing* 19 (2): 153–154.
Foucault, Michel. 1997. Self-writing. In *Michel Foucault, ethics: Subjectivity and truth*, ed. Paul Rabinow (Essential works of Michel Foucault 1), 207–222. New York: The New Press.
Friedrich, Markus. 2007. Communication and bureaucracy in the early modern Society of Jesus. *Schweizerische Zeitschrift für Religions- und Kulturgeschichte* 101: 49–75.
Gillum, Richard F. 2013. From papyrus to the electronic tablet: A brief history of the clinical medical record with lessons for the digital age. *American Journal of Medicine* 126: 853–857.
Hart, Curtis W., and M. Div. 2010. Present at the creation: The clinical pastoral movement and the origins of the dialogue between religion and psychiatry. *Journal of Religion and Health* 49: 536–546.
Hendrick, Burton J. 1914. Team work in healing the sick. *The World's Work* 2: 410–417.

Hunt, Linda M., Hannah S. Bell, Allison M. Baker, and Heather A. Howard. 2017. Electronic
    health records and the disappearing patient. *Medical Anthropology Quarterly* 31 (3): 403–421.
    https://doi.org/10.1111/maq.12375.
Myers-Shirk, Susan E. 2009. Helping the good shepherd. In *Pastoral counselors in a psychothera-
    peutic culture 1925–1975*. Baltimore: Johns Hopkins University Press.
Schutz, Roger. 1974. Kampf und Kontemplation. In *Auf der Suche nach Gemeinschaft mit allen*.
    Freiburg i.Br: Herder.
Vande Creek, Larry and Arthur M. Lucas. 2012 [2001]. The discipline for pastoral care giving. In
    Foundations for outcome oriented Chaplaincy. London: Routledge.
Worcester, Elwood, Samuel McComb, and Isador H. Coriat. 1908. *Religion and medicine: The
    moral control of nervous disorders*. New York: Moffat, Yard & Company.

Chaplaincy Documentation in a Large US Health System

Brent Peery

Chaplaincy documentation practices in the United States have evolved over time. Variation in practice still remains. However, the trend in the profession is toward the expectation that chaplains will document their care. There is also increased expectation regarding the content of that documentation. Demonstrating competency in clear and effective documentation is required for board certification and is identified among the expected standards of practice for the profession (Association for Clinical Pastoral Education et al. 2016; Association of Professional Chaplains Quality Committee 2015).

I have been writing and teaching on the subject since 2008. The content of this paper was developed in response to consistently strong energy, interest, and concern I have experienced from professional chaplains on the subject for the last decade. This growing interest seems to be driven by increased concern for the spiritual needs of patients and families within healthcare paired with the large number of healthcare organizations shifting to electronic medical records (EMRs). Though I have given this topic considerable thought, research, and practice, I do not pretend to have all the answers.

In this paper, I will share a little of the history and current practice of chaplaincy documentation within the Memorial Hermann Health System. This will include a discussion of the who, what, where, when, how, and why of chaplaincy documentation. Like many other chaplaincy departments, but by no means all, we take our documentation seriously. We try to document our care in a manner that bears witness to the holistic humanity of our care recipients. We also want to describe our care in a way that helps others understand the important ways professional chaplains contribute to the well-being of others.

B. Peery (✉)
Memorial Hermann Health System, Houston, TX, USA

21

1   History

The Memorial Hospital System originated in 1907, and the Hermann Hospital was opened in 1925. The two merged in 1997 to become the Memorial Hermann Health System. As of this writing, Memorial Hermann is comprised of 15 acute care hospitals, 2 physical rehab hospitals, 1 addiction treatment hospital, 1 hospice, 1 retirement home, and over 250 outpatient facilities in Houston, Texas, and the surrounding area. We have 42 professional chaplains, 11 chaplain residents, 5 administrative professionals, and 42 PRN (as needed) chaplains who serve in these entities. I have worked at Memorial Hermann for over 17 years as a chaplain resident, staff chaplain, chaplain manager, chaplain director, and currently as the vice president for chaplaincy.

I have spent my entire chaplaincy career in a setting where chaplaincy documentation has been ahead of the curve of national trends. Both professional chaplains and chaplain residents have documented their care in patients' medical records at Memorial Hermann and its predecessor organizations for over 45 years. With regard to this task, we are treated like any other healthcare professional on the interdisciplinary healthcare team. Chaplains documented exclusively in paper charts back in the 1970s and 1980s. Chaplains wrote in ink about their care in the patient's chart, which generally consisted of forms hole-punched and collated in a plastic binder. However, beginning in the late 1980s, Memorial Hermann began the slow transition to EMR. For over fifteen years we have almost exclusively documented our care in the EMR. Memorial Hermann's chaplains are required by current organizational policy to document their care for patients and families in our EMR software.

As a continuation of our emphasis on outcome-oriented chaplaincy[1] (OOC), which began in 2005, during 2006 and 2007, the professional chaplains of Memorial Hermann developed a uniform structure for chaplaincy documentation. Before that time, there was great variation in what Memorial Hermann chaplains documented about the care we provided. We came to believe a consistent approach to documentation across our system might help us be more intentional about our work and therefore improve our care. We also thought it would help other healthcare professionals understand better and potentially appreciate our work more. We reviewed the published literature on chaplaincy documentation at that time. In addition, we contacted a variety of hospitals throughout the United States to educate ourselves on existing chaplaincy charting practice. Then, we reviewed various models and discussed for months what we thought would be best documentation practice for us as chaplains. We finally identified what came to be called the Memorial Hermann Chaplaincy Documentation Model. It consisted of five parts:

(1) Reason for visit (Why is the chaplain involved?)

---

[1] "[Outcome-oriented chaplaincy] is a method of chaplaincy care that emphasizes achieving, describing, measuring, and improving outcomes that result from a chaplain's work. Its primary components include chaplaincy assessment, chaplaincy interventions, and chaplaincy outcomes" (Peery 2012a, b, 343).

(2) Interventions (What did the chaplain do to help the care recipient?)
(3) Outcomes (How did the care recipient respond? What difference did the chaplain's interventions make?)
(4) Assessment (How would the chaplain summarize this care recipient's current spiritual/emotional/relational state to the rest of the interdisciplinary healthcare team?)
(5) Plan (What does the chaplain intend to do further or recommend to the interdisciplinary healthcare team?)[2]

In the years since, many other professional chaplains from around the United States have adopted or adapted the Memorial Hermann Chaplaincy Documentation Model for their own practice.

In 2008, we designed and built a custom chaplaincy documentation template into our EMR software, based on our model. In 2013, we revised and expanded the content of our electronic template. The resulting new template was built into the software in 2014.

I learned some lessons from the experience of building and redesigning these templates. I share them here for those chaplains who may engage in a similar process. The first lesson I learned was to *know or learn what your priorities are* in this project. This will help you communicate clearly with the technical professionals with whom you will likely collaborate to build a custom documentation template.

Our task force identified four priorities in the most recent redesign of our template. The first priority was the ease of use of the template for entering data. We wanted to make it as easy as possible for chaplains to clearly document their care, in recognition of heavy workloads. The second priority was the clarity of the resulting chart notes. We wanted the data output to be clear to others who would read our notes. Our third priority was we wanted the template to be useful for chaplaincy education. We wanted the template to contribute to an action-reflection-action model of professional development. We designed it to help chaplains think through the content and efficacy of their work. A key part of this is the inclusion of extensive pre-supplied content. We envisioned this to be a design feature of particular value to chaplain residents and inexperienced chaplains. We also envisioned a template that helped other healthcare professionals more thoroughly understand the care provided by professional chaplains. Our final major priority was to build a template with the potential to use it to extract data for future research about our work. We had already tried and failed to get help in extracting modest data from our previous template. However, we were optimistic that the day would come when we could get that sort of analytics help. We wanted our template to be ready when the time came. The volume of pre-supplied content also contributed to this priority.

Another lesson I learned from our template design and build process was to *know or learn your limitations*. In our process, we encountered both software and institutional limitations. We had to make some compromises on our ideal design because

---

the software was not able to do what we preferred. We also encountered limitations because our institution had limited resources they could allocate to help us with our build. Despite those limitations that prevented us from getting everything we wanted, the end product of our most recent template has proven to be a valuable tool in our work for over five years.

Value the process was the final major lesson learned in our two documentation template designs and builds. From start to finish, each took about twelve to fifteen months. We work in a large organization and change can take a while. Along the way, we saw benefit in each of these from our discussions with each other about our work and how we describe that work to others. We debated about the wording of interventions and outcomes. We dialogued about our assessment process and descriptions. I think those discussions helped us to be better chaplains, apart from any product they produced.

2 General Guidelines

2.1 Who Reads Our Documentation?

There are a variety of persons who could potentially read our documentation. Most will fall into one of the following groups. These are listed in the order of likelihood to read according to our experience:

(1) Other chaplains – As a part of our chaplaincy assessment and care, we read other chaplains' documentation who have previously provided care to a patient/family and incorporate their insights.
(2) Other healthcare professionals – Nurses, physicians, social workers, and many other professionals utilize our expertise as recorded in our documentation to inform their own care for a patient/family.
(3) The patient and/or family – The patient has a legal right to obtain a copy of his/her medical records. The patient, or his/her surrogates, may read our documentation there.
(4) Members of the justice system – Occasionally a copy of a patient's medical record is subpoenaed by representatives of the civil or criminal justice system as part of their work. Chaplains have been asked to give testimony in legal proceedings based on the content of their documentation.

I have found it helpful to imagine representatives from all these groups looking over my shoulder as I document my care. We should document in a manner that communicates appropriately to all of these potential audiences.

2.2 What Do We Document?

There seem to be two broad schools of thought related to chaplaincy documentation. There are those that favor a minimalist approach and those that favor a more comprehensive approach. Because of increasing professionalism within chaplaincy, the minimalist school is shrinking. I have a clear bias toward the comprehensive school.

The minimalist school stresses documenting the bare minimum. A definition of minimal is "of the least possible; minimum or smallest" (Collins English Dictionary). Those who embrace a minimalist approach believe chaplains should document little or nothing. They argue that what we do cannot be described. Minimalists often identify confidentiality as the supreme value for guiding our documentation; documenting little leaves very little risk for violating confidentiality. An example of minimalist documentations would be "Provided chaplaincy care." When sociologist Wendy Cadge studied 19 chaplaincy departments in the United States for her book *Paging God: Religion in the Halls of Medicine*, she categorized those departments in three groups according to their level of professionalism. She assessed each department as being professional, transitional, or traditional. She identified these minimalist "I was here" chaplaincy documentation notes as characteristic of the traditional (least professional) departments (Cadge 2012, 114–121, 124, 139).

What are some the factors that commonly contribute to minimalist chaplaincy documentation? As was mentioned, a strong fear of violating confidentiality is often an issue. Because of where they trained and have worked, some chaplains never learned anything more than minimal documentation. For others, laziness is a factor. A few chaplains choose a minimalist approach because they resist any form of accountability for their work.

The comprehensive school seeks a more thorough approach to documentation. A definition of comprehensive is "of broad scope or content; including all or much" (Collins English Dictionary). Those who embrace a comprehensive approach believe chaplains should document like other healthcare professionals. Though we acknowledge there are aspects of chaplaincy care that defy description, we affirm that much of what we do can be described. For us, care is the supreme value for guiding our documentation. We are concerned that patients and families receive the best holistic care possible and believe thorough chaplaincy documentation will contribute to that goal. As we will discuss below, a comprehensive approach involves more than minimalist documentation and values concise communication.

The profession is increasingly moving toward a comprehensive approach to chaplaincy documentation. There has been some support for this approach dating back to at least 1940. In that year, Russell Dicks, chaplain at Presbyterian Hospital in Chicago, included detailed documentation in the patient's medical record among minimum recommended standards for chaplaincy (Dicks 1940). Every standards of practice document produced by the Association of Professional Chaplains since 2010 has included a standard for documentation. The most recent one is *Standards of Practice for Professional Chaplains*. Standard 3 states, "Documentation of Care:

The chaplain documents in the appropriate recording structure information relevant to the care recipient's well-being" (Association of Professional Chaplains Committee on Quality 2015).

Memorial Hermann chaplains describe our care through the five sections of the Memorial Hermann Chaplaincy Documentation Model: reason for visit, interventions, outcomes, assessment, and plan. Our primary areas of education, skill, and experience are:

(1) Spiritual
(2) Emotional
(3) Relational

Our documentation focuses mostly on these three areas of our professional expertise. Our chart notes should reflect what we assess through our senses – saw, heard, smelt, or felt – and what we did (interventions) related to these aspects of the care recipient's experience.

Because we are committed to holistic interdisciplinary healthcare, we also incorporate knowledge of the patient/family's biomedical needs, hopes, and resources into our care. These issues frequently impact a person's spiritual, emotional, and relational well-being. However, we rely upon other healthcare professionals with greater expertise in those areas to provide biomedical documentation. Chaplains are cautioned to refrain from documenting biomedical content. When referencing biomedical issues relevant to our chaplaincy care, an external source for that information should be documented (i.e., "Per RN's report, pt [patient] received a new diagnosis of diabetes today").

In general, chaplains should refrain from including any hearsay in their documentation. Among other potential reasons, in an effort to make sense of a patient's illness or injury, it is not unusual for first responders, family, healthcare providers, and sometimes the media to tell the story of how the patient came to be hospitalized. Very often early versions of those stories are revised with the passage of time and a more thorough collection of facts. For example, it is not generally appropriate for a chaplain to document "Pt was admitted for a self-inflicted gunshot wound to the head." This would very rarely be information the chaplain knows firsthand.

Important note: The Joint Commission (TJC) is a major accrediting organization for American hospitals. When the TJC does an onsite accreditation survey of a hospital, they do patient medical record "tracers" to evaluate the quality of care provided. These involve reading/tracing a patient's medical record from admission until the present. From a chaplaincy perspective, they expect our notes to tell the story of our care, including our assessment, interventions, and outcomes. Also, of particular concern for them is when any healthcare professional indicates a plan of care for the patient and family. In such cases, they want to see further documentation indicating the plan was implemented. When we indicate in our documentation a plan to follow up, we make sure we *do* follow up and document that care.

2.3 When Do We Document?

We document after every chaplaincy visit. As the saying around healthcare goes, "If it's not in the chart, it did not happen!" Sometimes we also document after unsuccessful attempts to visit. The latter practice can demonstrate responsiveness to referrals, even though circumstances may have prevented the visit (i.e., patient is away from their room for a procedure). It can also help facilitate communication within multi-staff chaplaincy departments.

It is best practice to document as soon after the visit as is practical. There will be times when demands are so intense that a chaplain is not able to document for a few hours. However, there are some benefits from charting after every visit or two, when possible. First, it yields more accurate notes. After the passage of time and multiple encounters, important details of our care can begin to fade from our minds. Second, it can facilitate chaplain self-care. This can be intense and draining work. The discipline of timely documentation can help the chaplain take a break and clear his/her mind from the previous visit before going on to the next. Third, documenting in the clinical setting between visits can facilitate staff care. It often creates opportunities for informal conversation at the nurses' station that lead to important caring opportunities for other healthcare professionals who need our support.

2.4 Where Do We Document?

Memorial Hermann chaplain documentation is recorded in a section of the patient's EMR where notes from other professionals on the interdisciplinary healthcare team also appear. This demonstrates the value of both holistic patient care and the valuable contributions chaplains make to that process.

2.5 Why Do We Document?

First and foremost, chaplains document because we are healthcare professionals. Healthcare professionals document their work. This is primarily to promote interdisciplinary communication and teamwork and optimize care. There are also legal and regulatory reasons why healthcare professionals must document their care.

There are other compelling reasons for chaplaincy documentation. We value holistic care for patients and families. Our notes are testimony that the patient is not just a body; he/she is a person with spirituality, thoughts, emotions, and relationships. Our documentation can help other healthcare professionals understand and respond helpfully to patient/family spiritual, emotional, and relational needs, hopes, and resources. Lastly, our documentation can help other healthcare professionals better understand chaplaincy care and partner with us more appropriately in caring for patients and families.

2.6 How Do We Document?

We document with *clarity*. Our work and the human beings for whom we care are very complex. However, we endeavor to clearly communicate both in our documentation.

We document with *conciseness*. As Gordon Hilsman writes, "Nobody reads long chart notes" (Hilsman 2017, 206). Clarity and brevity are related. Art Lucas contended that the ability to describe our work succinctly is evidence of clarity (VandeCreek and Lucas 2001, 19). This is difficult. It is a skill that gets better with practice.

We document with *care*. Our primary motive for all of our work should be care and concern for the other. This includes the work of documentation. We want to document enough to insure patients and families receive the best possible care. The Golden Rule applies. Document in a manner you would want for yourself or your loved one if the roles were reversed.

We document with *confidentiality*. "The chaplain respects the confidentiality of information from all sources, including the care recipient, legal, or organizational records, and other care providers in accordance with federal and state laws, regulations, and rules" (Association of Professional Chaplains Committee on Quality 2015, Standard 6, Confidentiality). The relationship between a chaplain and a patient/family is one of sacred trust. Our documentation should be respectful of the trust others give to us. There are, however, limits to confidentiality. In general, those limits are crossed when we learn someone has or will do harm to self or others. In such cases, we need to appropriately disclose such information. We also disclose other information to the patient's treatment team we believe will enhance that patient's care.

On the subject of confidentiality, like other US healthcare professionals, we are subject to patient privacy laws like the Health Insurance Portability and Accountability Act (HIPAA). This law was passed, in part, to make sure healthcare professionals respected the confidentiality of a patient's medical records. When we first access a patient's EMR, we are prompted by the software to identify our relationship with that patient (i.e., chaplain, chaplain resident/intern). We only access the EMR of patients when doing so is necessary for our work. Once in a patient's medical record, we only access the parts of that EMR necessary to provide spiritual, emotional, and relational support to that patient and his/her family. Chaplains should access a patient's EMR with professional integrity and a very high degree of respect for the privacy of those we serve. There is an enduring electronic record of all of our activity within a patient's EMR.

3   The MH Chaplaincy Documentation Template

The Memorial Hermann chaplaincy documentation template is designed to be a tool that helps chaplains clearly describe the care they have provided to patients and families. The Chaplain Visit form consists of five pages, corresponding with the five parts of the MH Chaplaincy Documentation Model (reason for visit, interventions, outcomes, assessment, and plan). On each of the five pages, there are both pre-supplied (click-to-select) content and free text boxes into which we can type original content. We can use either of these means of entering content or both in combination. MH chaplains may use their judgment and preferences to guide them to the best means to clearly communicate their care (The "What Do We Document?" and "How Do We Document?" sections above provide additional guidance). Other considerations include:

(1) New chaplain residents and interns may want to limit themselves mostly to the click-to-select content for a few weeks until they feel more comfortable and competent with documentation.
(2) The amount of content in our documentation is directly related to the duration and complexity of our chaplaincy care. A relatively brief encounter in which no chaplaincy care needs are assessed would generally yield a short chart note. A longer helping encounter, in which the need for more significant interventions and outcomes is assessed, should result in a more comprehensive chart note.

3.1   Significant Other (SO) Designation Form

There is also an optional SO Designation form. This form facilitates identifying SOs related to a particular patient and who are recipients of chaplaincy care. Common SO roles could include husband, wife, boyfriend, girlfriend, domestic partner, father, mother, daughter, son, brother, sister, friend, etc. These roles will be imported into all subsequent Chaplain Visit forms for this patient until a different SO Designation is completed for the patient. These roles can also be edited or entered in the Chaplain Visit form. Therefore, the SO Designation form is optional. It was developed so the chaplain would not have to identify SOs every time they document care to a particular patient/family (Fig. 1).

3.2   Reason for Visit

The Reason for Visit page includes a box to indicate the amount of time in minutes (rounded to the nearest 0.25 hours) spent providing chaplaincy care during this encounter. While this information is not directly related to the reason for the visit, it is valuable information to record about our care. Our chaplain design team chose to

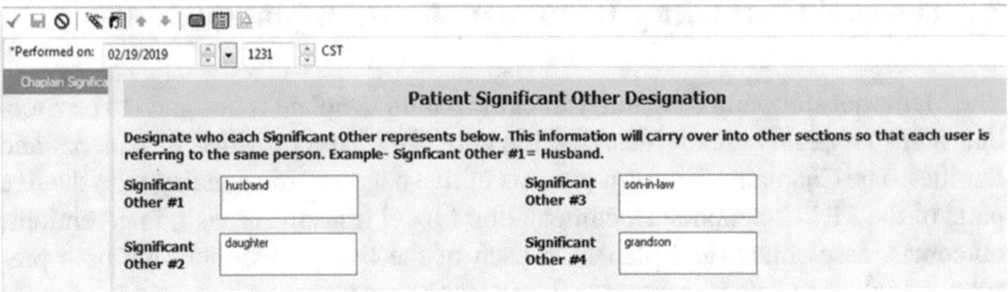

Fig. 1 Significant Other Designation Form. (Reproduced with permission from Memorial Hermann Health System. Copyright © 2013 Memorial Hermann. All rights reserved)

place the input for this information here because it seemed like the best among the available options. The second box contains several common click-to-select reasons for a chaplaincy visit. Note that selecting the "Referral from" option in the Reason for Visit box opens the third "Referral from" box. This box contains some common sources of referral. (Selecting "Other" in any box enables you to add an option that better communicates.) There is an "Additional Reasons for Visit" free text box into which you can type other information to convey the reason for the chaplaincy encounter. The "Patient Religious Preference" box imports the religious preference identified for the patient at admission. We cannot directly alter this information in the patient's EMR. However, we can confirm it is correct in our assessment. If it is inaccurate, we can contact the admissions department to get the information corrected (Fig. 2).

3.3 Interventions

When we expanded the pre-supplied list of common chaplaincy interventions from 34 to 44 interventions during the 2013 revision of this template, we decided to divide the interventions' list into five categories (empowerment, exploration, collaboration, relationship building, and ritual). This was intended to both broadly describe the kinds of interventions we provide and to reduce the length of the list a chaplain reads through when utilizing click-to-select content. All of the content entered on this page is action-oriented. It describes what the chaplain did to help the care recipient. For definitions and examples of each of the listed interventions, see the glossary below (Fig. 3).

Performed on: 02/19/2019   1236   CST

Chaplaincy Reason
Chaplaincy Interventions
Chaplaincy Outcomes
Chaplaincy Assessment
Chaplaincy Plan

Chaplaincy Reason

Time Spent

○ 15	○ 105
○ 30	○ 120
○ 45	○ 120+
○ 60	○ Other
○ 75	
○ 90	

Reason for Visit

☐ Admission request   ☐ Ethics consult   ☐ New diagnosis
☐ Advance directive request   ☐ Family meeting   ☐ Referral from
☐ Change in condition   ☐ Fetal demise   ☐ Rounds visit
☐ Chapel contact   ☐ Follow up visit   ☐ Support group
☐ Code/rapid response   ☐ Initial visit   ☐ Other
☐ End of life   ☐ IDT rounds

Referral from

☐ Case manager   ☐ Friend   ☐ Physical therapist   ☐ Other
☐ Chaplain   ☐ Ministry volunteer   ☐ Physician
☐ Child life   ☐ Nurse   ☐ Respiratory therapist
☐ Counselor   ☐ Occupational therapist   ☐ Social Worker
☐ Faith community   ☐ Patient   ☐ Speech therapist
☐ Family   ☐ Patient care assistant   ☐ Supportive/Palliative care

Additional Reasons for Visit

Patient Religious Preference (READ ONLY)

Catholic

Fig. 2 Reason for Visit Form. (Reproduced with permission from Memorial Hermann Health System. Copyright © 2013 Memorial Hermann. All rights reserved)

3.4 Outcomes

Art Lucas wrote mostly about outcomes in the future tense, i.e., "desired contributing outcomes" (VandeCreek and Lucas 2001, 18–21). We might also call these "chaplaincy care goals." In our documentation model, we would include goals in the plan section. We use the outcomes section to describe the observable differences our care made for the care recipient during the encounter just completed. As such, they are worded in the past tense. In describing outcomes, we acknowledge there may be significant outcomes that cannot be sensed or described. However, our experience has been that focusing on discernable outcomes helps us evaluate and improve our care. It also helps others better understand our care (For definitions and examples of each of the listed outcomes, see the glossary (Fig. 4)).

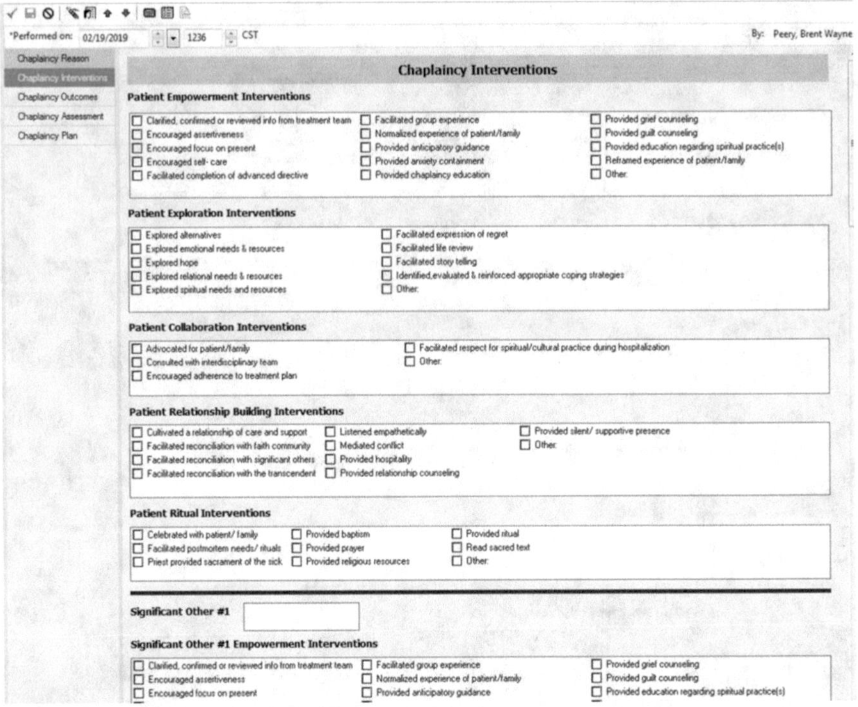

Fig. 3 Chaplaincy Interventions Form. (Reproduced with permission from Memorial Hermann Health System. Copyright © 2013 Memorial Hermann. All rights reserved)

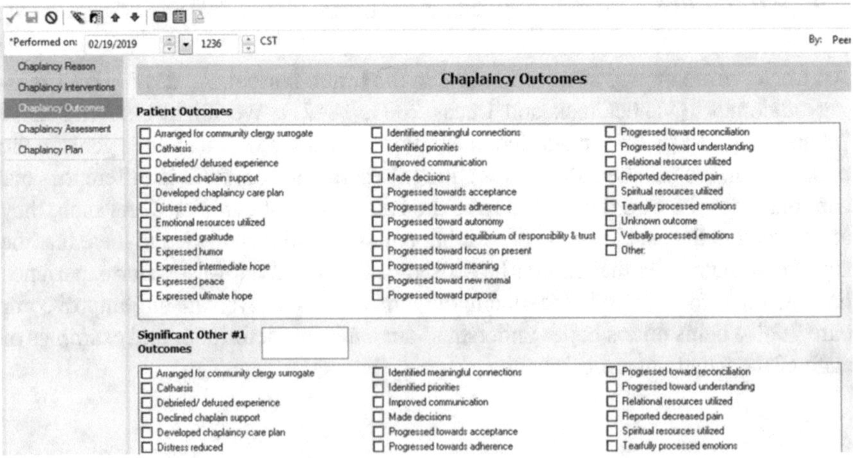

Fig. 4 Chaplaincy Outcomes Form. (Reproduced with permission from Memorial Hermann Health System. Copyright © 2013 Memorial Hermann. All rights reserved)

3.5 Assessment

Our goal in the assessment section is to succinctly summarize and communicate the current spiritual, emotional, and relational state of the recipients of our care. We are often asked about our documentation model, "Why do you document the assessment after the interventions? Don't you assess before you intervene?" It is true that we often assess before we intervene. However, chaplaincy assessment is more complex than such a simple linear conceptualization would illustrate. Our assessment is a dynamic process. It often begins before we even enter the room with the other. This early assessment work could be based on many factors such as what we were told by a referral source, what we read in the patient's EMR, where the patient is located, etc. We hold those early assessments loosely. As our caring encounters progress, we move fluidly between assessment, interventions, and outcomes. Sometimes early assessments need to be modified or discarded based on additional information. Sometimes the spiritual, emotional, or relational state of the other varies throughout the helping encounter. In light of this complexity, we asked ourselves, "What should we document?" Our answer is reflected in the first sentence in this paragraph. We decided this is the most helpful information to include. Further, we decided this was the best place in the documentation model to communicate it.[3] The assessment page in our documentation template is built around:

(1) Spiritual needs, hopes, and resources
(2) Emotional needs, hopes, and resources
(3) Relational needs, hopes, and resources

Our emphasis on assessing needs, hopes, and resources is rooted in the work of Art Lucas. He cautioned chaplains against accepting the primary pathological focus of medicine. He warned against the tendency to only see our care recipients in terms of their needs. Lucas encouraged chaplains to also assess the person's hopes and resources (VandeCreek and Lucas 2001, 8).

The "Patient/Significant Other Needs & Hopes" box (see screenshot below) is known as an ultra-grid in our EMR software. To access the click-to-select content for spiritual, emotional, or relational needs (first three columns), we click on the box under the appropriate heading and to the right of the person we are assessing. This opens a pop-up box with a pre-supplied list of common needs.[4] Chaplains can select any of the listed common needs assessed with the care recipient or type in any additional identified needs. The right two columns are for typing hopes the care recipient

---

[3] The MH Chaplaincy Documentation Model is a communication strategy. It is not necessarily a chronological retelling of all that occurred in the helping encounter.

[4] These lists include as follows: *Spiritual needs* – belief issues, guilt, lack of gratitude, lack of meaning, lack of peace, lack of purpose, lack of self-care, misinformation, providence issues, and theodicy issues. *Emotional needs* – anger, anxiety, betrayal, confusion, despair, fatigue, fear, frustration, grief, loneliness, resentment, sadness, and shame. *Relational needs* – conflict, lack of assertiveness, lack of autonomy, lack of communication, lack of companionship, lack of relational skills, lack of responsibility, lack of trust, rejection, and unrealistic expectations.

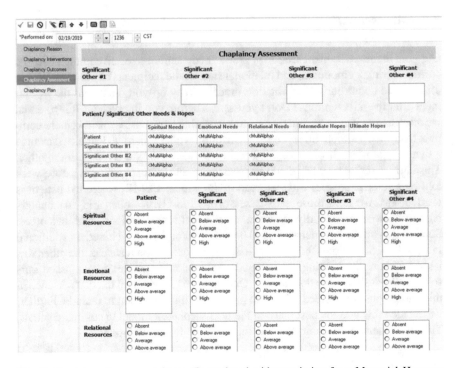

Fig. 5 Chaplaincy Assessment Form. (Reproduced with permission from Memorial Hermann Health System. Copyright © 2013 Memorial Hermann. All rights reserved)

mentions. Intermediate hopes are based in preferred outcomes (i.e., positive test result, restored function, cure, etc.). Ultimate hopes transcend any particular outcomes (i.e., trust in God, confidence in love and goodness, etc.) (VandeCreek and Lucas 2001, 14–15). Lester labeled these concepts as "finite" and "transfinite" hope (Lester 1995, 63–65) (Fig. 5).

The next portion of the assessment page enables the chaplain to evaluate the care recipient's spiritual, emotional, and relational resources. For each person, we can rate their resources in each category on a five-point scale from absent to high. If we rate all three categories of resources, a total resource score will appear in the appropriate box below. If we rate only one or two of the categories, our assessment of resources will be recorded without a total resource score (range of 1 to 15). The relative proportionality of needs, hopes, and resources is a key part of our assessment. High needs and hopes paired with limited resources could indicate greater need for chaplaincy care. Similarly high needs and hopes paired with strong resources likely indicate lower need for chaplaincy care. A caring, experienced, knowledgeable, and skilled chaplain is potentially a tremendous resource for others (Fig. 6).

The "Resources Identified" free text box allows us to specifically record any resources identified in our assessment.

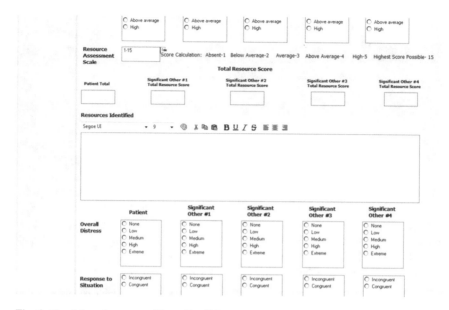

Fig. 6 Chaplaincy Assessment Form, Detail Resource Scale. (Reproduced with permission from Memorial Hermann Health System. Copyright © 2013 Memorial Hermann. All rights reserved)

The "Overall Distress" is a summary rating of the care recipient's suffering from none to extreme.

The "Response to Situation" has two choices. An "Incongruent" response is an indication the care recipient's overall distress is significantly higher or lower than what the chaplain has experienced from others in similar circumstances. Incongruence may be an indication of something that needs special chaplaincy attention and further assessment.

The "Subject Sated" free text box is used to record direct quotes from the care recipient relevant to the chaplaincy assessment (i.e., "I think I might be depressed" (Fig. 7)).

3.6 Plan

We indicate our plans for further care and recommendations to the interdisciplinary healthcare team on the plan page. The "Chaplaincy Plan" box contains four levels of our intent to follow up. Selecting "Will follow up" is an indication the care recipient definitely needs additional chaplaincy support. It also opens the "Follow up needed for" box in which the chaplain can indicate the areas of follow-up needs. Selecting "Follow up as circumstances allow" is an indication the care recipient could benefit from additional chaplaincy support if it can be arranged. Selecting

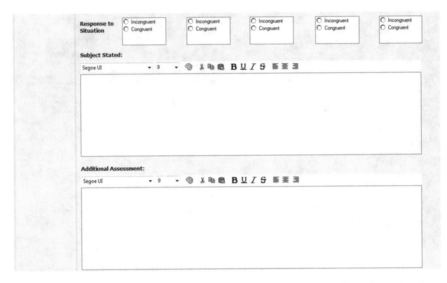

Fig. 7 Chaplaincy Assessment Form, Free Text Boxes. (Reproduced with permission from Memorial Hermann Health System. Copyright © 2013 Memorial Hermann. All rights reserved)

"Follow PRN" is an indication of no present need for additional chaplaincy support while acknowledging the situation is such that future needs may arise before discharge. Selecting "No follow up warranted at this time" is an indication there are no current or anticipated chaplaincy needs. Selecting "Will make referral to" opens the "Referral Needed to" box to indicate sources of referral help the chaplain will seek for the care recipient (Fig. 8).

4 Sample Chart Note

Our completed chart notes are more streamlined. They contain only the selected or typed content. All of the other items available in the extensive data input template are not imported into the final note. Below is an example of what one looks like.

Result type: Chaplaincy Note
Result date: 04/29/2018 11:22
Result status: Auth (Verified)
Result title: Chaplaincy Visit
Performed by: Chaplain, Charlie on 04/29/2018 11:22
Signed by: Chaplain, Charlie on 04/29/2018 11:22
Encounter info: HH HERMANN, Inpatient, 11/17/2017 –

Final Report *
Chaplaincy Visit Entered On: 04/29/2018 12:13
Performed On: 04/29/2018 11:22 by Chaplain, Charlie

Fig. 8 Chaplaincy plan form. (Reproduced with permission from Memorial Hermann Health System. Copyright © 2013 Memorial Hermann. All rights reserved(

Chaplaincy Reason

Chaplaincy Time Spent: 45
Chaplaincy Reason for Visit: Initial visit, Referral from, Rounds visit
Chaplain Referral Source: Nurse
CH Additional Visit Info: RN (Becky) told me Pt [abbreviation for "patient"] was "having a hard day" and asked for me to check on her.
CH Religious Preference: Methodist

Chaplaincy Interventions

Empowerment Interventions: Encouraged focus on present, Normalized experience of patient/family, Provided anticipatory guidance, Provided anxiety containment, Provided chaplaincy education, Provided education regarding spiritual practice(s)
Exploration Interventions: Explored emotional needs and resources, Explored hope, Explored relational needs and resources, Explored spiritual needs and resources, Facilitated storytelling, Identified, evaluated, and reinforced appropriate coping strategies
Collaboration Interventions: Consulted with interdisciplinary team, Encouraged adherence to treatment plan
Relationship Building Interventions: Cultivated a relationship of care and support, Listened empathetically

Ritual Interventions: Provided prayer
Significant Other #1 Title: husband
Empowerment Interventions SO1: Provided chaplaincy education
Significant Other #2 Title: daughter
Significant Other #3 Title: son-in-law
Significant Other #4 Title: grandson

Additional Interventions: Provided instruction re meditative prayer as an anxiety containment strategy

Chaplaincy Outcomes

Patient Outcomes: Debriefed/defused experience, Distress reduced, Emotional resources utilized, Expressed gratitude, Expressed intermediate hope, Expressed ultimate hope, Progressed toward focus on present, Relational resources utilized, Spiritual resources utilized, Tearfully processed emotions, Verbally processed emotions

Significant Other #1 Title: husband
Significant Other #1 Outcomes: Unknown outcome
Significant Other #2 Title: daughter
Significant Other #3 Title: son-in-law
Significant Other #4 Title: grandson

Chaplaincy Additional Outcomes: Pt shared her medical narrative. She identified and processed the spiritual and emotional sequelae. Reported anxiety decreased. Breathing slowed and became more regular. Pt smiled more.

Chaplaincy Assessment

Significant Other #1 Title: husband
Significant Other #2 Title: daughter
Significant Other #3 Title: son-in-law
Significant Other #4 Title: grandson

Pt Spiritual Resources: Above average
Pt Emotional Resources: Average
Pt Relational Resources: Above average
SO1 Relational Resources: Above average
Resource Assessment Scale: 1–15
Pt Resource Score: 11
CH Helpful Resources Identified: Pt articulated strong, long-tenured, and supportive personal Christian faith. She seems emotionally strong and mature. She mentioned her husband, adult children, and faith community (Grace United Methodist in Pleasant Prairie) as sources of good relational support.

Patient Overall Distress: Medium
SO1 Overall Distress: Low
Patient Congruency Response: Congruent
SO1 Response: Congruent

Subject Stated: Patient: "I know God will take care of me no matter what. He has brought me through so much. I am blessed!"

Additional Assessment: Husband left room after brief introduction. Pt mentioned normal mixed emotions re prospect of potentially curative brain surgery. She expressed moderate anxiety re unknown. She mentioned fears of not being a suitable surgical candidate and of possible lack of efficacy or of unintended consequences of surgery (lost function). She also expressed the strong desires to have the surgery and be sz [abbreviation for "seizure"] free for the first time in decades. Her anxiety seemed lessened after chaplaincy care. Per her report and chaplain's observations, she benefitted from chaplaincy support.

Chaplaincy Plan

Chaplaincy Follow- up: Will follow up, Will make referral

Follow-Up For: Spiritual support, Emotional support

Chaplain Referral To: Clergy/Faith community, Professional chaplain

Chaplaincy Additional Plan: Because Pt is scheduled for sleep deprivation tonight, will make referral to night chaplain (Allan Jenkins) for follow up. Per pt's request, will contact her pastor to request a hospital visit.

5   Chaplaincy Screening Process

In 2014 we implemented a new chaplaincy screening process in our EMR software. The purpose of this process is to identify patients/families with potential spiritual or religious struggle, as well as those who would like to receive chaplaincy support. After a thorough literature review and consulting with other healthcare organizations around the nation regarding best practice, we settled upon the Rush Protocol as the model for our screening process. It was originally developed by the chaplains at Rush University Medical Center in Chicago. At the time, it was the most thoroughly researched and validated screening tool for potential spiritual or religious struggle (cf. Fitchett and Risk 2009, 1–12, and see below for a diagram of the resulting Memorial Hermann Chaplaincy Screening Process). This screening process is a part of the nurse admission history completed in the EMR by a nurse caring for the patient/family. Patients/families are asked two to three questions (the number depends upon the answers they provide). A task is automatically generated in our department task list whenever a patient/family answers a question in a manner that indicates possible spiritual or religious struggle or when he/she requests chaplaincy support (Fig. 9).

Memorial Hermann Chaplaincy Screening Process

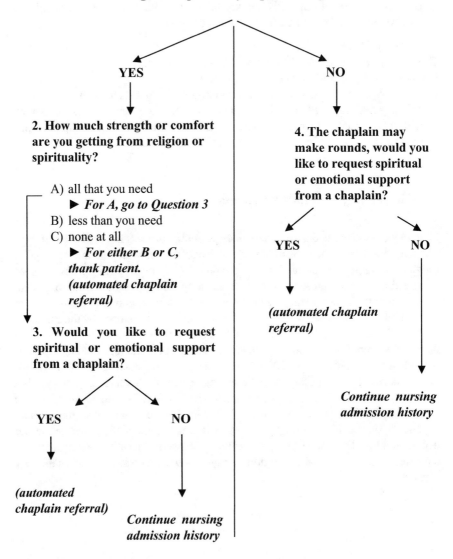

Introductory statement:
Many patients and families have religious or spiritual beliefs that give them
strength, comfort, and contribute to health.

**1. As you cope with your condition/illness/injury,
is religion or spirituality important to you?**

YES

NO

**2. How much strength or comfort
are you getting from religion or
spirituality?**

A) all that you need
  ▶ *For A, go to Question 3*
B) less than you need
C) none at all
  ▶ *For either B or C,
  thank patient.
  (automated chaplain
  referral)*

**3. Would you like to request
spiritual or emotional support
from a chaplain?**

YES

NO

*(automated
chaplain referral)*

*Continue nursing
admission history*

**4. The chaplain may
make rounds, would you
like to request spiritual
or emotional support
from a chaplain?**

YES

NO

*(automated chaplain
referral)*

*Continue nursing
admission history*

Fig. 9 Memorial Hermann Chaplaincy Screening Process (Adapted from The Rush Protocol in
Fitchett and Risk 2009. Screening for spiritual struggle, with permission from SAGE Publications.
*JournalofPastoralCare&Counseling*63(1–2):1–12.https://doi.org/10.1177/154230500906300104;
Copyright © 2009 SAGE Publications. All rights reserved)

6 Glossary

Interventions (44)
Empowerment

- **Clarified, confirmed, or reviewed information from treatment team** – chaplain helps patient/family understand information related to patient's treatment plan.

  Example: Chaplain assists spouse in recalling unfamiliar diagnostic term used by physician.

- **Encouraged assertiveness** – chaplain seeks to improve the care recipient's skills in expressing to others his/her concerns, thoughts, emotions, or rights.

  Example: Chaplain encourages patient to voice to her physician her disagreement with her treatment plan.

- **Encouraged focus on present** – chaplain helps the care recipient to experience the benefits of attention to the present in contrast to inordinate or unhealthy focus on the past or future.

  Example: Chaplain recognizes catastrophic thinking and anxiety about possible future events and suggests greater consideration of gifts and challenges of today.

- **Encouraged self-care** – chaplain encourages the care recipient to engage in basic activities that nourish and sustain spiritual, emotional, relational, and physical health.

  Example: Chaplain learns that parent of patient has neither slept nor eaten in over 24 hours and recommends both.

- **Facilitated completion of advance directive** – chaplain helps the care recipient consider options and complete advance directive document(s).

  Example: Chaplain helps patient evaluate and decide who among her family would make the best medical power of attorney for her.

- **Facilitated group experience** – chaplain leads a group meeting intended to contribute to the well-being of participants.

  Example: Chaplain leads support group for brain cancer patients and families.

- **Normalized experience of patient/family** – chaplain reassures the care recipient that new and distressing reactions are common to others with similar experiences and not pathological.

  Example: Husband says, "I'm losing my mind. I cannot remember anything the doctor told me about her condition." Chaplain responds, "Most people find it hard to think and remember under this kind of stress."

- **Provided anticipatory guidance** – chaplain informs or makes recommendations to the care recipient about anticipated experiences.

  Example: Chaplain describes common symptoms of approaching death to prevent distress among family gathered at terminal patient's bedside.

- **Provided anxiety containment** – chaplain helps the care recipient to strengthen their ability to respond constructively to unspecified threat(s).

  Example: Chaplain guides patient's partner through progressive relaxation exercise.

- **Provided chaplaincy education** – chaplain gives the care recipient information about how professional chaplains can help or how to access chaplaincy support.

  Example: Chaplain explains, "Chaplains are here for spiritual and emotional support for patients and families from diverse faith perspectives or no particular faith perspective."

- **Provided grief counseling** – chaplain helps the care recipient heal from the pain of loss.

  Example: Chaplain listens empathically to son's painful account of all he will miss after his mother's death.

- **Provided guilt counseling** – chaplain helps the care recipient understand and respond beneficially to emotions of guilt or shame.

  Example: Chaplain helps patient identify ways he can realign his behavior with his personal values.

- **Provided education regarding spiritual practice(s)** – chaplain teaches the care recipient beneficial spiritual practices.

  Example: Chaplain teaches patient a meditative prayer technique to aid in pain management.

- **Reframed experience of patient/family** – chaplain helps the care recipient consider his/her experience from a different and potentially more beneficial perspective.

  Example: Chaplain reflects, "I wonder if it was simply an oversight born of distraction?" in response to a description of another's offensive behavior.

Exploration
- **Explored alternatives** – chaplain helps the care recipient carefully consider his/her options.

  Example: Chaplain leads patient to verbalize treatment alternatives along with the pros and cons of each.

- **Explored emotional needs and resources** – chaplain helps the care recipient identify, evaluate, or gain insight into his/her emotions along with their current and potential contributions to his/her well-being.

  Example: Chaplain helps father identify the hurt and fear that underlies his expressed anger.

- **Explored hope** – chaplain helps the care recipient identify, understand, or rely upon sources of intermediate and ultimate hope.

  Example: Spouse says, "We haven't lost hope." Chaplain asks, "Upon what or whom does your hope rest?"

- **Explored relational needs and resources** – chaplain helps the care recipient identify, evaluate, or gain insight into his/her relationships along with their current and potential contributions to his/her well-being.

  Example: Chaplain helps patient to reflect upon parameters of control and trust in her significant relationships to achieve healthy equilibrium.

- **Explored spiritual needs and resources** – chaplain helps the care recipient identify, evaluate, or gain insight into his/her meaningful connections.[5]

  Example: Chaplain listens attentively and reflectively as patient shares his beliefs and the strength they provide.

- **Facilitated expression of regret** – chaplain provides a safe space for the care recipient to express regret over his/her thoughts, words, actions, or failure to act.

  Example: Chaplain carefully listens as a teenage patient confesses the recent mistakes she has made.

- **Facilitated life review** – chaplain helps the care recipient tell stories from his/her life to bring forth beneficial actions or attitudes.

  Example: Chaplain asks a patient with terminal illness, "What relationships have meant the most to you through the years?"

- **Facilitated storytelling** – chaplain helps the care recipient communicate his/her medical or life narrative.

  Example: Chaplain asks, "How did you end up here in the hospital?"

- **Identified, evaluated, and reinforced appropriate coping strategies** – chaplain elicits from the care recipient his/her strategies for confronting personal challenges and assesses their helpfulness.

---

[5] "Spirituality is the aspect of humanity that refers to the way individuals seek and express meaning and purpose and the way they experience their connectedness to the moment, to self, to others, to nature, and to the significant or sacred" (Puchalski et al. 2009, 885–904).

Example: Chaplain asks, "Have you identified anything that has helped you deal with anxiety in the past?"

Collaboration

- **Advocated for patient/family** – chaplain speaks, writes, or acts on behalf of the care recipient's needs, especially when institutional or authority structures have not satisfactorily addressed those needs.

  Example: Chaplain appeals to another healthcare professional to alter patient's treatment plan to accommodate patient/family spiritual, emotional, and/or relational needs.

- **Consulted with interdisciplinary team** – chaplain consults with other healthcare professional(s) to acquire or provide information intended to optimize care for patient/family.

  Example: Chaplain participates in multidisciplinary rounds meeting.

- **Encouraged adherence to treatment plan** – chaplain encourages the care recipient to engage suggested activity or inactivity prescribed by treatment team.

  Example: Chaplain affirms to a severely depressed patient the value of regularly taking his psycho-pharmaceutical medications.

Facilitated respect for spiritual/cultural practice during hospitalization – chaplain helps the care recipient maintain meaningful practice while in the hospital.

  Example: Chaplain arranges for halal diet for Muslim patient.

Relationship Building

- **Cultivated a relationship of care and support** – chaplain introduces self and begins to establish a caring relationship through which the care recipient is helped.

  Example: "Hello, Mr. Simpson, I am Chaplain Mary. I stopped by to see how you are doing today."

- **Facilitated reconciliation with faith community** – chaplain helps the care recipient resolve estrangement from his/her faith community.

  Example: Upon hearing patient's grief from having "drifted away" from his synagogue, with patient's permission, chaplain calls rabbi to arrange visit at hospital.

- **Facilitated reconciliation with significant others** – chaplain helps the care recipient restore human relationships damaged by past action, misunderstanding, or neglect.

  Example: Chaplain coaches dying patient through a process of apologizing to his children for past neglect.

- **Facilitated reconciliation with the transcendent** – chaplain helps the care recipient restore a transcendent relationship damaged by past action, misunderstanding, or neglect.

  Example: Chaplain helps sister of patient seek forgiveness from and express love to God.

- **Listened empathically** – chaplain listens in a manner that conveys to the care recipient cognitive and affective perception and understanding of the care recipient's experience.

  Example: Chaplain says, "This sounds like it has been extremely difficult for you."

- **Mediated conflict** – chaplain helps care recipients to de-escalate or resolve conflict.

  Example: After vocal disagreement between ICU patient's ex-wives, chaplain helps family agree upon visitation schedule acceptable to all.

- **Provided hospitality** – chaplain helps the care recipient through small acts of service.

  Example: Chaplain provides ice water and coffee for family as they await an update about their loved one's surgery.

- **Provided relationship counseling** – chaplain helps the care recipient experience more satisfying relationships.

  Example: Chaplain suggests a strategy for parenting a toddler.

- **Provided silent and supportive presence** – chaplain accompanies the care recipient while nonverbally communicating empathy and support.

  Example: Chaplain gently touches a bereaved husband's shoulder and supplies tissue as he weeps over his wife's death.

Ritual
- **Celebrated with patient/family** – chaplain joins the other in expressing positive emotions resulting from a preferred circumstance.

  Example: Chaplain rejoices with patient/family about a positive medical test result.

- **Facilitated postmortem needs/rituals** – chaplain helps the care recipient care for the body of loved one according to his/her beliefs.

  Example: Chaplain arranges for Buddhist priest to perform postmortem ritual for deceased patient and insures with healthcare team that the body will remain undisturbed for 8 hours.

- **Priest provided sacrament of the sick** – Roman Catholic priest provides sacrament of the sick (often referred to by outdated term "last rites") for care recipient.

Example: Chaplain arranges for family's priest to provide sacrament of the sick.

- **Provided baptism** – chaplain performs rite of Christian baptism for care recipient.

  Example: Chaplain baptizes terminally ill infant.

- **Provided prayer** – chaplain prays (aloud, silently, with others, or alone) for the care recipient according to the care recipient's preference and faith tradition.

  Example: Chaplain joins a Roman Catholic family in reciting the "Our Father."

- **Provided religious resources** – chaplain supplies literature, objects, or personnel intended to facilitate the care recipient's religious practice.

  Example: Chaplain provides flameless candles for Jewish family Sabbath ceremony.

- **Provided ritual** – chaplain supplies rite or ceremony meaningful to the care recipient.

  Example: Chaplain conducts naming and blessing ceremony for parents of still-born child.

- **Read sacred text** – chaplain reads aloud written material revered by the care recipient.

  Example: Chaplain reads a Psalm at patient's bedside.

Outcomes (32)

- **Arranged for community clergy surrogate** – after confirming that patient meets the criteria[6] for a clergy surrogate decision maker, chaplain enlists clergy surrogate.

  Example: Chaplain reviews patients' medical record, consults with treatment team, identifies clergy surrogate, and facilitates communication between surrogate and physician.

- **Catharsis** – a sudden expression of previously restrained emotions resulting in a reduction of distress

  Example: Patient angrily shares a litany of complaints about her hospital experience which culminates in a fit of sobbing and concludes with "I feel better."

- **Debriefed/defused experience** – an oral recounting of events and associated reactions, often resulting in decreased emotional tension and/or clarified cognition.

  Example: Mother tells the story of the automobile accident that resulted in her child's hospitalization.

---

[6] Texas Health and Safety Code Chap. 313: Consent to Medical Treatment Act

- **Declined chaplain support** – choosing not to receive chaplaincy care.

  Example: After chaplain introduction, patient's husband says, "We're all fine here. But, thanks for stopping by."

- **Developed chaplaincy care plan** – chaplain and care recipient collaborate to identify desired chaplaincy outcomes and anticipated interventions intended to realize them.

  Example: Chaplain and patient with an intrauterine fetal demise agree on a plan for a naming and blessing ceremony after she delivers her baby.

- **Distress reduced** – unpleasant or disturbing emotions are decreased.

  Example: Patient's primary facial expression transitions from fearful to happy.

- **Emotional resources utilized** – care recipient is able to make use of sources of inner emotional strength to cope with present needs.

  *Example: Patient draws upon her athletic experience and learned ability to delay gratificati*on in order to avoid discouragement in physical therapy.

- **Expressed gratitude** – care recipient communicates thankfulness.

  Example: Son mentions thanks that his mother's cancer was discovered before metastasis.

- **Expressed humor** – care recipient employs humor.

  Example: Family mixes funny accounts from his life with their tears around patient's deathbed.

- **Expressed intermediate hope** – care recipient articulates hope based on preferred future outcomes.

  Example: Patient mentions her desire to regain enough agility to be able to play on the floor with her grandchildren.

- **Expressed peace** – care recipient communicates intrapersonal and/or interpersonal serenity.

  Example: Patient describes a lack of inner conflict about his chosen treatment plan.

- **Expressed ultimate hope** – care recipient articulates hope not based on preferred future outcomes.

  Example: Daughter states, "I would rather my father is able to live a few more years. But, whether he lives or dies, I know he will be safe in God's loving care."

- **Identified meaningful connections** – care recipient names relationships with persons, places, activities, or ideas that provide meaning and/or purpose.

  Example: Patient discusses how much she values three lifelong friends.

- **Identified priorities** – care recipient describes what is of greatest importance.

  Example: Patient expresses his preference to remain alert and able to communicate over complete pain control.

- **Improved communication** – care recipients are able to more clearly exchange messages.

  Example: Mother and physician are each better able to understand each other's perspective.

- **Made decisions** – care recipient decides among alternatives.

  Example: Patient chooses among possible discharge options.

- **Progressed toward acceptance** – care recipient moves toward acquiescence of unwanted realities.

  Example: Paraplegic is able to acknowledge a meaningful life without walking.

- **Progressed toward adherence** – care recipient increasingly follows his/her medical treatment plan.

  Example: Depressed patient agrees to attend weekly psychotherapy sessions and take antidepressant medication.

- **Progressed toward autonomy** – care recipient moves toward freedom and self-governance.

  Example: Abused wife makes a plan to seek the assistance of a local women's shelter after discharge from hospital.

- **Progressed toward equilibrium of responsibility and trust** – care recipient is more able to take appropriate personal responsibility while trusting others to fulfill their responsibilities.

  Example: Mother is able to acknowledge both that she took reasonable precautions to ensure her child's safety and another's carelessness caused his hospitalization.

- **Progressed toward focus on present** – care recipient increasingly experiences the benefits of attention to the present in contrast to inordinate, or unhealthy, focus on the past or future.

  Example: Patient is able to self-correct when his thoughts and conversations are focused too much on the past or future.

- **Progressed toward meaning** – care recipient is better able to identify and describe significance.

  Example: Patient states, "I guess my suffering was meant to help me slow down and love more."

- **Progressed toward new normal** – care recipient increasingly accepts new circumstances and finds satisfying ways to live within them.

  Example: Paraplegic patient joins and enjoys a wheelchair basketball league.

- **Progressed toward purpose** – care recipient is better able to identify and describe desired ends or consequences from his/her life.

  Example: Patient comments, "I want to be remembered as a decent man who loved people, especially my family and friends."

- **Progressed toward reconciliation** – care recipient experiences a reduction in relational conflict or estrangement.

  Example: Patient's daughter decides to call her brother to whom she has not spoken in several years.

- **Progressed toward understanding** – care recipient more clearly perceives reality.

  Example: Wife articulates accurately her husband's condition, prognosis, and treatment plan.

- **Relational resources utilized** – care recipient is able to make use of social support network to cope with present needs.

  Example: Mother of patient accepts neighbor's offer to care for her other children.

- **Reported decreased pain** – care recipient communicates lessened spiritual, emotional, relational, or physical pain.

  Example: After expressing secret feelings of guilt patient states, "It feels good to get that off my chest."

- **Spiritual resources utilized** – care recipient is able to make use of sources of spiritual strength to cope with present needs.

  Example: Patient expresses verbally and nonverbally confidence in the care of a loving higher power.

- **Tearfully processed emotions** – care recipient expresses emotions through the means of tears.

  Example: Father intensely sobs as he holds the body of his stillborn baby.

- **Unknown outcome** – chaplaincy care produced unidentified results.

  Example: Dementia patient communicates no discernible reaction to chaplaincy interventions.

- **Verbally processed emotions** – care recipient expresses emotions through the means of words.

  Example: Trauma patient talks through her feelings about the drunken driver that caused injury to her and her family.

Commentary

Simon Peng-Keller

1. A Milestone in Digital Recording Spiritual Care

In order to understand why practices of recording spiritual care in EMR have been gaining currency in recent years, it is helpful to study paradigmatic cases. Brent Peery offers not only an illuminating example but also a model of best practice. It is taken from an institution which has an extraordinarily long history of chaplaincy records. The chapter can be seen as an exercise in affirmative genealogy: By telling the history of the current practice in the Memorial Hermann Health System in Houston, Peery explains which considerations, experiences, and decisions are behind it – historically as well as structurally. One might put the chapter's main thesis as follows: In order to offer the best possible spiritual care in the context of the constraints of current health systems, chaplains have a double task: first, they are required to develop a recording tool suitable for spiritual care in the framework of a specific institution; second, they need to train themselves in the use of the tool to record their work. Beyond any doubt, digital charting changes the daily practice of healthcare chaplains considerably. Charting is not only time-consuming: it implies restructuring one's work, framing one's perceptions, modifying one's communication. Peery doesn't ignore the challenges, but he emphasizes the advantages of this development. In my commentary I will focus, first, on the conceptual framework of the model described. In particular, I examine its relationship to the paradigm of outcome-oriented chaplaincy. With regard to possible implementations of the model, I analyze, second, the information provided by the chapter about the tool and the practice of recording in the Memorial Hermann Health System.

2. Recording Outcome-Oriented Chaplaincy

Three factors explain why recording spiritual care in the Memorial Hermann Health System has taken the elaborated form described in Peery's chapter: the long tradition of charting chaplaincy in medical records, the more recent emergence of EMR, and, not least, the paradigm of outcome-oriented chaplaincy. As we outline in the introduction, the latter was developed by Art Lucas in the 1990s at Barnes Jewish Hospital in Saint Louis, Missouri. Lucas was dissatisfied with the standard paradigm of Clinical Pastoral Education centered in recent decades on unintentional presence. His vision was to improve spiritual care by developing and implementing a more structured and more disciplined approach, influenced probably by his Methodist background. When planning this volume and inviting the contributors, we didn't realize that two of them were so deeply influenced by Lucas (cf. the contribution of Anne Vandenhoeck). This may not be a mere coincidence. There is a mutual affinity between this new paradigm of healthcare chaplaincy and the emergence of the EMR which provides a technical structure for the former. What Lucas developed with ink and paper fits perfectly with the new world of digital health. In the following, I shall consider this mutual affinity with regard to the paradigm case of Memorial Hermann.

The traces of Lucas are to be found on different levels. First of all, the structure is clearly outcome-oriented. Clear objectives and procedures are at stake: assessment, plan, intervention, outcomes. The MH model does not merely reflect a very structured process of spiritual care, rather it compels chaplains to structure their work in a predetermined manner. Peery states it clearly: "The MH chaplaincy documentation model is a communication strategy." Lucas's influence is also reflected in key concepts, even though some of them are also used elsewhere (e.g., the difference between intermediate and ultimate hope). Finally, the signature of the outcome-oriented paradigm is to be seen in the computing of "overall distress," a procedure for measuring the emergency (cf. Figure 6).

The MH chaplaincy documentation model may give rise to at least two sorts of qualms. The first has to do with language, the second with structure. Is the language used suitable for the practice it is supposed to describe and orientate? Or asked more pointedly: Are the outcomes of outcome-oriented language in line with the objectives of chaplaincy itself? Is it compatible with a spirituality which is formed by the logic of an unfathomable gift, not primarily by plans, goals, and success? It can hardly be denied: As language creates reality and as digital instruments (and their realities) form today's clinical practice, the concepts and tools for describing spiritual care have remarkable outcomes themselves. One could relativize the objection with the hint that the administrative language is to be found mainly on the level of the super-categories (assessment, plan, interventions, outcomes), while the subcategories remain the traditional ones (prayer, anointment of the sick, etc.). This may be one of the compromises necessary for interprofessional communication. Nevertheless, I cannot help but consider it a misclassification to subsume prayer under intervention.

The qualms with structure are connected to the question of the power of medical and administrative language. In counterbalancing those approaches to chaplaincy that overemphasize unintentional presence, Lucas provides an important inspiration for the development of spiritual care. My concern here is over-structuring, a problem inherent in all models and tools of documentation. In our workshop Brent Peery argued convincingly that through good training and experience chaplains may attain the ability to use the digital tool in a flexible manner. For the further development of chaplaincy worldwide, it would be useful to document such individual learning processes in recording spiritual care!

3. The Core of Spiritual Care?

A lot of training and experience is required to master a fine-grained digital tool for recording spiritual care. Only healthcare chaplains working with it permanently are able to acquire the skills and the habits needed. This ability, then, will separate board-certified healthcare chaplains from visiting ministers on the one hand and healthcare professionals on the other hand. It is supposed to facilitate interprofessional communication. But what role is to be played by physicians, nurses, and other care-givers in the interprofessional spiritual care epitomized by this tool? (Considering this aspect of the MH model, Michael and Tracy Balboni's critical remark came to my mind: "that professionalization of chaplains will undermine the role of [...] medical professionals in offering spiritual care" [Hostility to Hospitality. Spirituality and Professional Socialization within Medicine, Oxford 2019, 252–253]. I doubt, however, that the professionalization of healthcare chaplaincy must have the effect of diminishing the spiritual care specific to physicians, nurses, and other caregivers.) Are they to be mere readers? Or do they have their own tools for recording the spiritual care that forms part of their professional activities?

Learning to chart is becoming part of the basic training for healthcare chaplaincy, part of the process of its ongoing professionalization. The 44 interventions listed in the glossary are very revealing as to what these professionals are up to do besides charting. I read the list carefully against the background of my own experiences as a part-time chaplain in a palliative care unit. Most of the activities listed are part of my work and that of my colleagues. (Some further activities could be added, for example, singing or humming religious and nonreligious songs.) But it is the first time I have seen them listed in such a complete and orderly fashion. Without any doubt, well-structured tools for recording spiritual care foster reflection and awareness. Considering the entries, one might conclude that the main focus of a chaplain's work lays on psychosocial support. Most of the "interventions" could also be provided by a psychologist. The old question of the Clinical Pastoral Education, the relationship between pastoral and psychological counseling, reemerges here. Strikingly, psychologists are missing from the list of other professions eligible for referrals (cf. Figure 8).

4. Final Thoughts

What impressed me most when I became acquainted with MH's digital tool was its elasticity and user-friendliness. Chaplains have enough space for longer narrative entries, but they can confine themselves to a few clicks as well. In each case, the computer will summarize all entries neatly and calculate, if possible, the total distress. Developed by what Peery calls the "Comprehensive School," the tool could also be useful for partisans of the "Minimalist School" or other schools. With this remarkable offspring of years of intensive work and experience, Brent Peery and his colleagues have set a milestone in the development of the charting of spiritual care in EMR.

References

Association for Clinical Pastoral Education, Association of Professional Chaplains, Canadian Association for Spiritual Care, National Association of Catholic Chaplains, and Neshama: Association of Jewish Chaplains. 2016. *Common qualifications and competencies for professional chaplains.*

Association of Professional Chaplains Committee on Quality. 2015. *Standards of practice for professional chaplains.* Accessed on 2/15/19 at http://www.professionalchaplains.org/Files/professional_standards/standards_of_practice/Standards_of_Practice_for_Professional_Chaplains_102215.pdf.

Cadge, Wendy. 2012. *Paging god: Religion in the halls of medicine.* Chicago: The University of Chicago Press.

Comprehensive. n.d. *Collins English dictionary – complete & unabridged 10th edition.* Retrieved February 20, 2016 from Dictionary.com website http://dictionary.reference.com/browse/comprehensive.

Dicks, Russell. 1940. *Standards for the work of the chaplain in the general hospital.* Hospitals November.

Fitchett, George, and James Risk. 2009. Screening for spiritual struggle. *The Journal of Pastoral Care & Counseling* 63: 12.

Hilsman, Gordon. 2017. *Spiritual care in common terms: How chaplains can effectively describe the spiritual needs of patients in medical records.* Philadelphia: Jessica Kingsley Publishers.

Lester, Andrew. 1995. *Hope in pastoral care and counseling,* 63–65. Louisville: Westminster John Knox Press.

Minimal. n.d. *Collins English dictionary – Complete & Unabridged 10th Edition.* Retrieved February 20, 2016 from Dictionary.com website http://dictionary.reference.com/browse/minimal.

Peery, Brent. 2008. Chaplaincy charting: One healthcare system's model. *PlainViews* 5:8. May 21.

———. 2012a. Describing our magic. *PlainViews* 9:18. October 17.

———. 2012b. Outcome oriented chaplaincy: Intentional caring. In *Professional spiritual and pastoral care: A practical clergy and chaplain's handbook,* ed. S. Roberts, 342–361. Woodstock: SkyLight Paths Publishing.

Puchalski, Christina, et al. 2009. Improving the quality of spiritual care as a dimension of palliative care: The report of the consensus conference. *Journal of Palliative Medicine* 12 (10): 885–904.

Texas Health and Safety Code Chapter 313: Consent to Medical Treatment Act.

VandeCreek, Larry, and Art Lucas, eds. 2001. *The discipline for pastoral care giving: Foundations for outcome oriented chaplaincy.* Binghamton: The Haworth Press.

The Quebec Model of Recording Spiritual Care: Concepts and Guidelines

Bruno Bélanger, Line Beauregard, Mario Bélanger, and Chantal Bergeron

1 Background

In the health system, the user's social and health-care record is, so to speak, the container in which all relevant information about the professional services the user has required and received is kept. It is an essential tool, and one of its key functions is to promote communication between all those involved in the patient's treatment. In accordance with the *Organization and Management of Institutions Regulation/ Règlement sur l'organisation et l'administration des établissements* in Quebec, a note is written each time some form of professional action is taken;[1] it is the 'preferred instrument for demonstrating prudent and diligent professional conduct that meets ethical and civil obligations'.[2] The health-care record gives the various persons involved a better understanding of the user's situation whilst ensuring continuity and completeness of care and services.

In Quebec, spiritual care services receive their funding following the compilation of 'units of measurement'. The types of spiritual care intervention that qualify as units are listed in a document produced by the *Ministère de la Santé et des Services sociaux* (MSSS) in 2002. For each unit of measurement reported, there must be a

---

[1] These are articles 17 to 28 of the 'Act Respecting Health Services and Social Services, the Organization and Management of Institutions Regulation' (see no 53); the 'Règlements sur la tenue des dossiers de certains ordres professionnels' and, in a supplementary capacity, the 'Act on access governed by the content, use and access to files held by health institutions'. For example, Article 50 of the 'Organization and Management of Institutions Regulation' stipulates that 'every institution shall keep an individual record for each beneficiary who obtains services from it'. http:// legisquebec.gouv.qc.ca/en/ ShowDoc/cs/S-4.2; http://legisquebec.gouv.qc.ca/en/ShowDoc/cr/S-5,%20r.%205/

[2] La tenue de dossier, guide explicatif. Ordre des psychologues du Québec, January 2006, p. 3.

B. Bélanger (✉) · L. Beauregard · M. Bélanger · C. Bergeron
Centre Spiritualitésanté de la Capitale-Nationale., Quebec City, Canada

© The Author(s) 2020
S. Peng-Keller, D. Neuhold (eds.), *Charting Spiritual Care*,
https://doi.org/10.1007/978-3-030-47070-8_4

note on file.[3] From 2003 onwards, CSsanté[4] introduced the practice of 'note to file' into its institutions. At the time, this was a major step in the development of the profession and an indication of the importance that providing accountability in spiritual care activities would assume in the years to come.

In 2008 we wrote a guide to support spiritual care providers with this new task.[5] Over several years of using this guide, we have identified two weaknesses in our overall assessment process and in our methods of writing the note to file: (1) there was considerable variation in the way things were done in our own institutions; (2) the language used was sometimes too imprecise for other professionals. Moreover, more generally, the context in which our practice evolved had changed profoundly. As elsewhere in the world, the position of spiritual care professionals has changed over the course of recent decades. We have witnessed a major development: a *pastoral paradigm*, in which interventions took place in connection with a church or religious tradition, has moved towards a *biomedical paradigm* (Rumbold 2013).

In Quebec, this new situation can be seen in particular in the changes made in job titles by the Ministry of Health and Social Services: until the 1960s spiritual care professionals were known as *chaplains* (*aumôniers*), and then they became *pastoral facilitators* (*animateurs de pastorale*) and then, in 2011, *spiritual care providers* (SCPs) – *intervenants en soins spirituels* (*ISS*). Whilst integrating the richness and diversity of spiritual and religious traditions, the profession has had to state the purpose and objectives of intervention autonomously, i.e. without any reference to a particular church.[6] Under this new model, intervention practices tend to be carried out and assessed in a context of interdisciplinary collaboration, hence the need to develop language that is intelligible and meaningful to other members of the healthcare team.

Particularly following the change of the profession's name in 2011, work has been undertaken to further improve the practices associated with writing notes. We therefore worked on improving the practices involved in writing notes to file, initially developing an assessment tool, which was agreed upon by all SCPs at CSsanté. This gave rise to a further tool, *Repères pour l'évaluation en soins spirituels* (RESS) (Markers for Spiritual Care Assessment), on the basis of which a guide for writing notes to file was created.

---

[3] The evaluation note always corresponds to a unit of measurement, but not the intervention notes that follow.

[4] The mission of the 'Centre Spiritualitésanté de la Capitale-Nationale' (CSsanté) is to ensure planning, coordination, provision and assessment of quality clinical activities, in order to respond to the spiritual and religious needs of people who are hospitalized, in residential homes or receiving care at home. It thus brings under one umbrella all the spiritual care provided by health institutions and the social services of the city of Quebec and the surrounding districts of the Capitale-Nationale. We have a team of over 40 spiritual care coordinators working at 35 sites belonging to three institutions in the health and social services network.

[5] Centre de pastorale de la santé et des services sociaux. Guide de rédaction de notes au dossier en pastorale de la santé, 2008.

[6] In Quebec, the profession was closely linked to the Catholic Church; up until 2011, most professionals received their pastoral mandate from the church.

This document presents the key elements in the practice of writing notes to file commonly applied at CSsanté. There are two main parts: the first, conceptual in nature, presents the RESS tool and the spiritual vision on which it is based; the second, practical in nature, presents the two guides used to write notes, both the assessment/intervention note[7] generally used in the short term and the assessment/intervention note employed when users can no longer express themselves.[8]

2   Concepts

As mentioned above, the first part presents the RESS assessment tool and the spiritual vision which underlies it. This conceptual section also sets out how notes to file are written.

2.1   *Markers for Spiritual Care Assessment (RESS)*

The RESS tool was developed in a three-stage approach: (1) a literature review (2013–2014), which allowed us to identify the main elements which would form the basis of the first working draft; (2) a pilot of this first draft, carried out by CSsanté in the light of the SCPs' clinical experience[9] (2014–2015); and (3) validation of the draft tool, by conducting research designed to assess its applicability,[10] carried out with about 40 SCPs working outside of CSsanté (2016–2017). Figure 1 presents the tool in its current form.

When designing the tool, we used the notion of 'markers' (repères), which allow us to identify the main themes that can be raised with patients during an assessment meeting. The Larousse dictionary defines 'repère' as that which makes it possible to identify something in a whole, to locate something in time and space. Thus, when recalling and analysing a meeting, the SCPs can use the markers to help them identify the key elements of the encounter. The markers provided in our tool are four topics which may be addressed at the outset by the patients: beliefs and practices,

---

[7] Following the evaluation meeting, the SCP (ISS) will record the patient's score on a progress note sheet, if necessary.

[8] In our territory, this scheme is mainly used in our long-term centres, but it could be appropriate in the short-term work, particularly in geriatrics.

[9] At CSsanté we hold regular workshops on clinical work. A group of six or seven SCPs led by a clinical coordinator gets together to exchange experiences and know-how, with the aim of improving practices.

[10] Research entitled 'Développement d'un outil pour l'évaluation en soins spirituels' (Developing a spiritual care assessment tool) by Bruno Bélanger, Mario Bélanger, Chantal Bergeron, Line Beauregard and Guy Jobin, 2016–2017. In all, 43 SCPs from nine institutions in the Quebec health network participated in the project.

Markers for Spiritual Care Assessement
in response to suffering, illness and death

Markers can appear in different ways: well-being, illness and suffering.

Centre Spiritualitésanté de la Capitale-Nationale (CSSsanté) 2020

Fig. 1 Markers for Spiritual Care Assessment (RESS). (Reproduced with permission from Centre Spiritualitésanté de la Capitale-Nationale' (CSsanté). Copyright © 2019 CSsanté. All rights reserved)

hopes, relationships and values/commitments.[11] We were already familiar with the STIV tools,[12] developed and used in Switzerland in particular, and ST-VIAR,[13] developed as well as used in Quebec and taught as part of the initial training recognized by the l'Association des intervenants et intervenantes en soins spirituels du Québec (AIISSQ). As Fig. 1 shows, this tool distinguishes itself from STIV and ST-VIAR in envisaging meaning as a transversal element. In fact, our clinical experience has led us to consider meaning as a series of questions and reflections which generally emerge in close connection with the four key markers. We have also placed a circle at the centre of our diagram bearing the words 'transcendent experience'. This expression, borrowed from Louis Roy (2014), relates to a patient's capacity or potential to evoke indications of a transcendent experience.[14] In our diagram, the circle also represents what the spiritual care interventions aim to achieve.

---

[11] These four markers may be expressed by the patient in terms of spiritual suffering or of spiritual well-being.

[12] STIV: Sens, Transcendance, Identité et Valeurs

[13] ST-VIAR: Sens, Transcendance, Valeurs, Identité, Appartenance, Rituels

[14] We explain in more detail what we mean by 'transcendent experience' in the Spiritual Vision Underlying RESS section.

We pursued three objectives in developing the tool: (1) to provide the SCPs with a practical and flexible tool involving markers which would aid them in their assessment work; (2) to present in attractive and concise visual form our understanding of spirituality in the context of illness; and (3) to facilitate and harmonize the writing of notes to file.

2.2   Spiritual Vision Underlying RESS

The assessment tool derives from our spiritual vision and our understanding of different expressions of spirituality during illness. We present this vision in the different points below.

We adhere to the tripartite or ternary anthropological vision which sees human beings as comprising three dimensions: the physical, mental and spiritual. This position is inspired in particular by the writings of anthropologist Michel Fromaget (2007, 2008a, b, 2009) but can also be found amongst other writers: Ugeux (2001), Rosselet (2002), Zundel (2005), Kellen (2015), De Lubac (1990), de Hennezel (1997, cited in Fromaget, 2007), Bryson (2015), etc. Tripartite anthropology is closely related to the holistic vision (bio-psycho-socio and spiritual) of human beings proposed in the Quebec health network, in particular in the field of palliative care. In subscribing to the fact that a human being comprises three dimensions, we assume that a person suffering physically is also affected mentally and spiritually. In the same way, if a person is suffering spiritually, the other dimensions of their life are most likely affected as well.

2.2.1   Spiritual Dimension: Access to the World of the Essence

The spiritual dimension makes direct reference to the word 'spirit'. With reference to Maître Eckhart, Fromaget (2007) indicates in his work *Naître et mourir* (Birth and Death) that we cannot comprehend what the spirit is if we have not yet experienced it ourselves. Despite the difficulties involved in objectivizing the spirit, we decided to present this dimension as an 'openness' to the world of essences (Fromaget 2007)[15] which can be accessed via contemplation. Several authors (Fromaget 2007, 2008a, b, 2009), Kellen (2015), Zundel (2005) and Roy (2014) write about the possibility for all human beings, regardless of their beliefs, to have one or several spiritual or transcendent experiences.[16] Spiritual life is thus seen as a

---

[15] See also Rosselet and Collaud, writers who inspired CSsanté's approach.

[16] The etymology of the word 'transcendence' has two roots: 'trans' and 'ascendere', 'ascendere' means 'to rise' and 'trans' means 'across' or 'beyond'. According to Roy (2014, 222), the words 'transcendence' and 'self-transcendence' characterize this phenomenon as movement across and beyond human actions. [...] Transcendence is expressed through human actions and puts us in direct contact with the beyond. Roy (2014, 223) understands self-transcendence to be a situation in which human being begins to move outside of themselves when they pose questions regarding the

potential for experience or as a potential for transformation, which appeals to the freedom of each individual.[17] Pargament (2007) proposes the term 'search for the sacred' to talk about this spiritual dimension. The word 'sacred' is associated with concepts of God, the Divine, transcendent reality, etc. Spiritual life can thus be considered as potential and as a search or quest.

2.2.2   The Transcendent Experience

As mentioned above, the idea of the transcendent experience is key to our RESS model. This expression is borrowed directly from Louis Roy (2014), the title of one of his works being *Transcendent Experiences: Phenomenology and Critique*. We found that this approach was compatible with the literature which had inspired us up to that time. Roy describes a transcendent experience as:

> Il s'agit d'une appréhension – c'est-à-dire d'une sorte de conscience (awareness), de connaissance intuitive qui capte l'attention d'une personne ou d'un groupe parce qu'elle est véhiculée par une sorte de sentiment spécial. Le sentiment colore notre réponse à quelque chose qui apparaît immense. Lorsque nous prenons contact avec une quantité ou une qualité infinie, nous pouvons avoir l'impression que cette dimension déborde notre vie 'normale', qu'elle ne saurait être contenue artificiellement dans les limites familières et qu'elle commande donc un respect profond. (Roy 2014, 15).

This idea of an experience which goes beyond normal life is referred to by other authors but characterized in a different way. Roy (2014) gives several examples: spiritual experience, peak experience, cosmic consciousness, religious experience, sign of transcendence, etc. The central circle of our diagram therefore designates both the possibility of discerning, in the patient's discourse, signs of a possible experience of transcendence and the intention of the intervention itself. The four basic markers can be 'heard' by other professionals (social worker, psychologist, nurse in particular), because they are like vestibules through which one must move in order to enter the person's inner universe. The ability to recognize and accompany the person by listening to the richest experiences of his or her life (what we call 'experience of transcendence') and the questions of meaning that are related to it is a skill that essentially falls within our field of intervention and our formation.

---

what, why and how of what they perceive. Such questions force us to search for relationships between things and their meaning. [...] Human beings do not only ask questions about meaning but also about truth and value.

[17]Zundel (2005, 65) explains: 'Heidegger asserts, with remarkable profundity, that a person's being resides in their ability to exist. This essentially amounts to saying that man given to himself by his physical birth remains open, incomplete, unfinished. Whilst other beings, animal, vegetable and mineral are "in-sistent" – that is to say, opaque to themselves and closed to themselves – human beings are "ek-sistent", destined to turn outwards, to choose at each moment between potential ways of being, i.e. to choose oneself, as Sartre puts it'.

2.2.3 An Experience Recognized by Its Fruits

For many authors, the spiritual or transcendent experience can be seen in the fruits that it bears. Fromaget (2008a) highlights this: 'Just one solution remains, therefore: to appreciate the depth of the experience by its impact, by the fruits that it bears. And the idea of fruit is an excellent one here, provided that the source of the fruit is unambiguous' (Fromaget 2008a, 5). What are these fruits? Fromaget is basically talking about the fruit of the spirit as described in Paul's Letter to the Galatians 5, 'love, joy, peace, forbearance, kindness, goodness, faithfulness, gentleness and self-control'. Roy (2014), meanwhile, refers to other fruits: the disappearance of fear and the establishment of profound peace; the acceptance of death and the absence of distress; serenity, peace which lasts for hours and days; and power and creativity. Even if, during a spiritual experience, the outward world changes in appearance, this does not mean that the world itself changes but that it is perceived at a different level (Fromaget 2008b, 6). Didier Caenepeel (2017) in his study of the work by Éric-Emmanuel Schmitt, *Oscar et la Dame Rose*, gave a good example of this reality in a conference at CSsanté in Quebec in 2017. He referred to the possible fruits arising from the support given to a young boy in palliative care. The reality of approaching death remains, but the child's perception of the absurdity of this reality changes, allowing him to experience a certain degree of peace and to live his final days better. We had these fruits or signs in mind when we placed at the centre of our RESS model 'the transcendent experience/signs in which it is manifest', examples of things which outwardly reveal a possible experience of this type.

2.2.4 Crisis as a Path to a Transcendent Experience

We recognize three main paths which can lead to a transcendent experience: the emotion of love, wonder in the face of beauty and a third, which is crucial in the field of health, the crisis or approach of death (Zundel 2005; Kellen 2015; Fromaget 2008a,b). In the hospital environment, people often experience a difficult moment, and this is a moment in which those providing support can ideally listen out for the possibility of such an experience and be aware of it. In fact, '[…] major transitions and life crises […] reveal the deepest dimension of life. Similarly, loss, accident, injury, trauma and disaster can push people to confront the finitude and precariousness of their lives and direct them to look beyond their immediate worlds' (Pargament 2007, 66). Indeed, considering the crisis patients might experience, spiritual support in the health network is particularly appropriate. The crisis often creates a nodal point which can affect the inner life of the sick person. Marin (2013, 15), talking about the ordeal of illness, describes the crisis well: 'for some, it is the experience of radical change, the terrible discovery that there is a strangeness at the very heart of intimacy. It requires nothing less than a redefinition of the self'. RESS was developed with this in mind. Our vision of spirituality is closely linked to the ordeal of illness and the suffering it can engender.

2.2.5   A Process…

When talking about the spiritual dimension of human beings, many works focus on the meaning and ultimate aim of human existence. The spiritual life is therefore considered not only as all the elements which can be objectified but also as a process or dynamic experience (Rumbold 2013; Waaijman 2006a). Bryson remarks on this issue: 'The other point about spirituality is that it evolves as a person's life experiences accrue. Spirituality is a process rather than an event' (Bryson 2015, 92). Indeed, a person can experience intense moments of suffering or even of spiritual distress, which then disappear and make way for moments of peace and hope (transcendent experience), or their illness may destroy their sense of peace and tranquillity, leading them to see the ordeal they are experiencing as something absurd… The crisis often gives rise to change in a person's inner life (Waaijman 2006b). In providing support at this time, SCPs are witnesses to this process. The assessment model reflects this dynamic process, signaling that the intended aim is always to achieve peace and hope (circle at the centre of the illustration); however, during the development of the illness, the patient may in fact oscillate between feelings of well-being and spiritual suffering. We considered this aspect by focusing on the displacement – the movement, represented by the gradation of colours – between the four basic landmarks and the experience of transcendence through meaning. Spiritual life is presented as a process of transformation (sometimes slow, sometimes fast, sometimes surprising), of change and searching (Pargament 2007), in particular in the way of seeing or perceiving what occurs, of experiencing the crisis.[18]

2.2.6   'Tracking the Theological'

Underlying our model is the conviction that the highest spiritual realities are expressed in the ordinary words a person uses[19] and that the SCP essentially 'hears' the way in which each patient makes sense of the crisis they are experiencing in their everyday speech (each person makes sense of their experience on the basis of a number of sources) (Rumbold 2013). The expression 'tracking the theological' (Dumas 2010, 200) corresponds well to this situation. It refers 'to the subtle presence of God at the heart of the world, a presence-absence that is impossible to define or categorize. The theological is elusive, but can nonetheless be found in the nooks and crannies of everyday life'. It is precisely this idea which underlies our vision of spirituality and our assessment tool. The four key markers on which the tool is based merge visually with meaning and the transcendent experience to show that spiritual life can generally be seen in the questions and

---

[18] 'Spirituality involves more than a substantive content area. It is not a static, frozen set of beliefs or practices. It is instead a process of searching, a search for the sacred' (Pargament 2007, 52).

[19] Rosselet (2002, 6) says in this regard: 'It seems from experience that the "loftiest" questions are very deeply rooted in the simplest everyday life'.

reflections (the experience) that emerge from the patient's daily life and that these can attain profoundly spiritual depths.

3 Note-writing Guides

This part gives a practical description of the two guides,[20] each containing a model plus explanations for writing notes: both the assessment/intervention note generally used in the short term and the assessment/intervention note employed when users can no longer express themselves.

3.1 *Assessment/Intervention Note*

The assessment/intervention note is structured directly on the RESS assessment tool shown in Fig. 1. We call this an assessment/intervention note because an assessment meeting often includes an intervention. Figure 2 shows a general model for writing this kind of note, i.e. the way in which it is set out.

Each of the points in the diagram in Fig. 2 is presented below as they appear on the patient evaluation form with an SCP.

3.1.1 Context of Assessment/Intervention

Reason for Request

In this section, the SCP indicates whether it is a presentation/evaluation visit or a visit following a reference. If it is a reference, the SCP indicates the reason. The SCP also records the receipt of consent from the user or family member (Fig. 3).

Patient's Condition

The SCP indicates here what he or she perceives of the patient's condition[21] and notes whether a relative or relatives are present (Fig. 4).

---

[20] A good assessment/intervention note must respect certain quality criteria; these are set out in Appendix A.

[21] In some of our institutions, information of this type is already recorded in the patient's electronic record and can be transferred directly to the SCP's notes.

| **1. Context of assessment/intervention** |
| **1.1 Reason for request** |
| 1.2 Patient's condition |
| 1.3 Sociodemographic data |
| **2. Exploring markers** |
| 2.1 Beliefs/practices |
| 2.2 Hopes |
| 2.3 Relationships |
| 2.4 Values/commitments |
| 2.5 Transcendent experience shared by the user |
| **3. Professional analysis/opinion** |
| 3.1 Identification of a support marker and an obstacle marker in the user's path |
| 3.2 Summary and professional opinion based on analysis model |
| **4. Intervention conducted (where appropriate)** |
| 4.1 Results |
| **5. Follow up and support plan** |
| 5.1 Follow up |
| 5.2 Support plan |
| **6. Consent to care** |

Fig. 2 Model for writing an assessment/intervention note with a patient. (Reproduced with permission from Centre Spiritualitésanté de la Capitale-Nationale' (CSsanté). Copyright © 2019 CSsanté. All rights reserved)

Date and time of the evaluation

0   Presentation/evaluation visit

0   Referred by:
     0   Physician   0   Interdisciplinary team
     0   Nurse   0   relative
     0   SCP colleague   0   Patient on his/her initiative
     0   Professional

0   Referred for:
     0   New diagnosis   0   Support
     0   Death   0   Terminal phase
     0   Intense sadness   0   Questions/discerning
     0   Rituals   0   Imminent death
     0   Needs related to relatives   0   Other

Consent of the user/relatives

0   yes
0   no

Fig. 3 Reason for request. (Reproduced with permission from Centre Spiritualitésanté de la Capitale-Nationale' (CSsanté). Copyright © 2019 CSsanté. All rights reserved)

We also document the sociodemographic data (Fig. 5)

State of consciousness		
0	Alert	0 Asleep
0	Agitated	0 Unconscious
0	Confused	0 Deceased

The interview took place
0 Patient only
0 In presence of a relative

Fig. 4 Patient's condition. (Reproduced with permission from Centre Spiritualitésanté de la Capitale-Nationale' (CSsanté). Copyright © 2019 CSsanté. All rights reserved)

Sex: 0 M   0 W

Age: _____

Civil status

0	Single		0	Married
0	Common-law partner		0	Separated
0	Divorced		0	Religious
0	Widowed			

Life situation

0	Lives alone			
0	Lives with a host family		0	Lives with one or more family members
0	Lives with children		0	Homeless
0	Lives in shared custody		0	Lives with his/her parents
0	Lives with a friend			

Family composition: _____ (optional text field)

_____

Religious or spiritual affiliation:

0	Buddhism		0	Judaism
0	Christianism		0	Islam
	0	Anglican	0	Traditional autochthonous spirituality
	0	Catholic	0	Secular spirituality
	0	Protestant	0	Atheism
0	Hinduism		0	Other: _____
				(text field)

Fig. 5 Sociodemographic data. (Reproduced with permission from Centre Spiritualitésanté de la Capitale-Nationale' (CSsanté). Copyright © 2019 CSsanté. All rights reserved)

3.1.2 Exploring Markers

In this section, the SCP briefly describes how they understand the situation on the basis of the four key markers.[22] This can be done in one or two sentences, summarizing what seemed significant and relevant to them during the meeting. In this section, the SCP may report facts, topics or themes addressed by the user, where appropriate recording some of their words directly, so the SCP can clearly demonstrate what they have understood of the patient's experience on the basis of the key markers. It is preferable not to record anything next to a marker if little was said regarding this subject. Besides the four key markers, in this section we have included the option of noting elements linked to what is referred to in the RESS as the transcendent experience (Roy 2014).[23] The ability to listen and then converse with a patient around such an experience is a fundamental part of his spiritual life. As indicated in the section Spiritual Vision Underlying RESS (cf. Sect. 2.2), the transcendent experience can be expressed especially when the patient recalls rich experiences in their life that have generated feelings of peace and tranquillity. We believe that referring to these experiences on the basis of the support marker is very important in professional analysis and writing.

Beliefs/Practices

- Adherence to multiform currents or ideas, often from particular cultures, situated in a continuum that can go as far as a transcendent faith that engages a whole life
- Practices and behaviours generated by beliefs
- Individual or collective manifestations

Hopes

- Ability/incapacity to project into the future (in the present life)
- What one can see ahead (possibilities/deadlines)

Relationships

- Relationships between people and their links and influences
- Impact of the disease on past and present relationships
- Major relational and emotional issues (love, conflicts, forgiveness, etc.)

---

[22] In exploring the markers, we suggest indicating the main elements arising from the four key markers and the transcendent experience. The 'meaning' marker, as seen in Fig. 1, is transversal, linking all others. We will look specifically at this reference point in Sect. 3, Analysis/Opinion.

[23] The transcendent experience is not easy to define objectively. It is considered in the section in which we discuss our vision of spirituality. Several works in the bibliography may provide greater explanation. See in particular Fromaget, Roy, Zundel.

Values/Commitments

– What mattered, what had weight in life, the ethics of the person
– The implementation of values in a particular field of commitment
– Values and commitments that persist or change

Transcendent Experience Shared by the User

– Emergence of a new being

  • Feeling of strength and courage, peace and communion, elevation, presence and liberation
  • Sensitivity to beauty and love
  • Disappearance of fear and distress
  • The feeling of being loved by God, the power to hope

– The experience of transcendence may manifest itself gradually in the person's history or suddenly and unexpectedly.

3.1.3 Professional Analysis/Opinion

In this section, the SCP can go through a series of steps which make it easier to identify the information in what they have heard and noted down. This important analytical stage helps the SCP to form a professional opinion about the patient's state and subsequently to draw up an intervention plan.

There are two parts to this section. Firstly, the SCP identifies a support marker and an obstacle marker[24] along the user's personal path. The SCP then offers a synthesis of his opinion using the concepts of spiritual well-being, discomfort and suffering.[25]

Identifying a Support Marker and an Obstacle Marker Along the User's Personal Path

The SCP, recalling the key aspects of the meeting, identifies one or more support marker and one or more obstacle maker. The support marker is that which seems to give the patient strength and reassurance during their illness. The obstacle marker is

---

[24] In some cases, there is no support marker: the patient's suffering is overwhelming. Alternatively, there is no obstacle marker: the patient seems to be bathing in a state of spiritual well-being.

[25] Document developed by the Spiritual Care Team of the CSSS Pierre-Boucher: Julie Bolduc, ISS, coordonnatrice professionnelle, Bruno Godbout, ISS, Ivan Marcil, ISS et Johanne Philipps, ISS (nov. 2014). Note that to maximize consistency with the type of vocabulary used in our text, we have replaced the word 'indicateur' with 'repère'.

that which seems to cause the patient the most suffering and difficulty during the ordeal they are undergoing (Fig. 6).

Summary and Professional Opinion Based on Analysis Model

The SCPs offer here a synthesis of their professional opinion using the three concepts of spiritual well-being, discomfort and suffering. In order to refine the note and enlighten the care team about their opinion, they may check off some meaningful elements within the selected field. For this step, the SCP can refer to Appendix B, which gives the opportunity to refine its analysis (Fig. 7).

Support Marker		Obstacle marker	
0	Beliefs and practices	0	Beliefs and practices
0	Hopes	0	Hopes
0	Relationships	0	Relationships
0	Values and commitments	0	Values and commitments
0	Transcendental experience *		

Fig. 6 Support/obstacle marker. (Reproduced with permission from Centre Spiritualitésanté de la Capitale-Nationale' (CSsanté). Copyright © 2019 CSsanté. All rights reserved)
*Transcendent experience is found only in the support markers since this marker implies a state of well-being and peace

0	**Spiritual well-being**
0	Seems at peace
0	Is hopeful
0	Is comfortable with existential questions
0	Is able to see meaning despite the situation (significance, direction, sensations)
0	**Spiritual discomfort**
0	Has questions or misunderstandings
0	Is on the way to inner peace
0	Needs to be comforted in his/her hope
0	Is in search of meaning (meaning, direction, sensations)
0	**Spiritual suffering**
0	Gives rise to profound existential questions
0	Has no hope
0	Cannot see meaning (significance, direction, sensations) in their situation

Fig. 7 Summary and professional opinion. (Reproduced with permission from Centre Spiritualitésanté de la Capitale-Nationale' (CSsanté). Copyright © 2019 CSsanté. All rights reserved)

3.1.4 Intervention Conducted (Where Appropriate)

Where appropriate, the SCPs will check one of the interventions. They also note any effects observed on the patient or family members (Fig. 8).

3.1.5 Follow-Up and Support Plan

In this section, the SCP records whether a follow-up meeting is necessary, depending on the focus of the intervention (Fig. 9).

Support Plan

If there is a follow-up, the SCP indicates the objective(s) of this possible intervention by checking or labelling its own objective(s) in the 'Other' sections (Fig. 10).

3.1.6 Consent to care

For ethical and legal reasons, the SCP confirms in this section that the user has consented to the meeting and to any possible follow-up work (Fig. 11).

0	Assistance relationship
0	Spiritual support
0	Ritual
0	Discernment or accompaniment of spiritual issues
0	Other
Results of the intervention: _____	

Fig. 8 Intervention conducted and results. (Reproduced with permission from Centre Spiritualitésanté de la Capitale-Nationale' (CSsanté). Copyright © 2019 CSsanté. All rights reserved)

0	Follow-up
0	Follow-up if possible
0	No follow-up for the moment
0	Referral to an SCP or other professional
0	Put in touch with a particular faith community

Fig. 9 Follow-up. (Reproduced with permission from Centre Spiritualitésanté de la Capitale-Nationale' (CSsanté) Copyright © 2019 CSsanté. All rights reserved)

Beliefs and practices

0   Discuss with the person the meaning of their beliefs and practices in relation to their
     experience.
0   Offer the person a ritual that allows him/her to reconnect with the richness of his/her inner
     life.
0   Other _____

Hopes

0   Support the person so that he or she can look to the future with greater serenity.
0   Other _____

Relationships

0   Accompany the person, in order to help him/her find peace despite the relational challenges
     imposed on him/her.
0   Other _____

Values and commitments

0   Accompany the person, in order to help him/her identify new values that emerge as a result
     of the experience he/she is going through.
0   Support the person in prioritizing the values that emerge as a result of the disease.
0   Other _____

Fig. 10 Support plan. (Reproduced with permission from Centre Spiritualitésanté de la Capitale-Nationale' (CSsanté) Copyright © 2019 CSsanté. All rights reserved)

0   The patient agrees with the proposed intervention plan.
0   The patient's clinical condition does not allow for agreement with the proposed intervention
     plan.
0   The patient disagrees with the intervention plan and the proposed recommendations.
0   The patient authorizes us to communicate with his relatives.
0   Family members agree with the proposed intervention plan.

Fig. 11 Consent. (Reproduced with permission from Centre Spiritualitésanté de la Capitale-Nationale' (CSsanté) Copyright © 2019 CSsanté. All rights reserved)

3.2 Note to File When Patients Can No Longer Express Themselves

The diagram in Fig. 2 can be used when the SCP is able to communicate directly with the patient. In some circumstances this is not possible because many patients can no longer express themselves as a result of dementia or some other disability (e.g. aphasia). In Quebec, such clients are often resident in a long-term care establishment (*centre d'hébergement de soins de longue durée [CHSLD]*) and pose a particular challenge for spiritual care provision. The support of a community of helpers and carers is necessary to recognize their communication difficulties. Indeed, sound knowledge of the patients' background will ensure that patients in long-term care establishments can live in accordance with their values, beliefs and the meaning they have given to their lives. Making an assessment and recording it

1. Context of the evaluation
1.1 Consent and expectations
1.2 Sources of information consulted
1.3 Diagnosis and condition of the patient
1.4 Religious or spiritual affiliation
2. Exploring markers
2.1 Beliefs/practices
2.2 Hopes
2.3 Relationships
2.4 Values/commitments
2.5 Transcendent experience
3. Professional analysis and opinion
3.1 Markers in support of the intervention
3.2 Obstacles related to each of the markers
4. Orientations/Follow-up/Plan
4.1 Follow-ups
4.2 Plan of intervention

Fig. 12 Diagram for writing notes with a relative (when a patient cannot express himself/herself). (Reproduced with permission from Centre Spiritualitésanté de la Capitale-Nationale' (CSsanté). Copyright © 2019 CSsanté. All rights reserved)

in a note to file is essential, but this can only be done with the help of a relative or close friend who provides the information necessary in the support process.[26]

Figure 12 shows a model for drawing up the note to file for patients who can no longer express themselves. This model is similar to that in Fig. 2, but the fact that the patient is known via an intermediary is taken into account throughout, and the note reflects this (Fig. 12).

Below we explain in detail each of the points in Fig. 12.

3.2.1 Context of the Evaluation

Figure 13 presents the elements that are considered in the context of evaluation for patients who can no longer express themselves.

---

[26] We have drawn up this guide for patients who can no longer express themselves, considering that for many of these patients, a relative or friend may be able to provide enough information to the SCP for them to be able to intervene appropriately. In other cases, the SCP can obtain the information from the patient record.

1.1 Consent and expectations

The evaluation was carried out with:
       0   Relative(s)
       0   Patient and
       Please specify: _____

Consent of next of kin
       0   yes
       0   no
       Please specify: _____

       Requests and expectations: _____

This patient has received unction   date: _____

1.2 Sources of information consulted

    0   Clinical Team   0   Volunteers
    0   Relatives   0   Other
    0   Dossier

Additional information and details:

_____

1.3 Diagnosis and condition of the patient

       Diagnosis: _____
       Condition of the patient

    0   Alert   0   Dozy
    0   Agitated   0   Unconscious
    0   Confused   0   Deceased
    0   Asleep

Précisez | Please specify:_____

1.4 Religious or spiritual affiliation

0   Buddhism
0   Judaism
0   Christianism   0   Islam
    0   Anglican   0   Traditional autochthonous spirituality
    0   Catholic   0   Secular spirituality
    0   Protestant   0   Atheism
0   Hinduism   0   Other: _____

Fig. 13 Context of the evaluation. (Reproduced with permission from Centre Spiritualitésanté de la Capitale-Nationale' (CSsanté). Copyright © 2019 CSsanté. All rights reserved)

3.2.2 Exploring Markers

In this section, the SCP describes what they hear during the meeting with the patient's close family member or friend (see Sect. 3.1.2).

3.2.3 Professional Analysis and Opinion

In this section, the SCP picks out one support marker and one obstacle marker (Fig. 14). This analysis is important as it forms the basis for determining which interventions should take place with the patient.[27] The SCP takes into account the support marker in order to maintain the resident in the continuity of his history (which made sense) by promoting moments of comfort, peace and joy. The obstacle marker is taken into account by the SCP in order to promote, despite this suffering (which did not make sense), trust, forgiveness, hope, etc.

3.2.4 Orientations/Follow-up/Plan

In this section the SCP indicates what he or she considers appropriate as future directions for this patient. If follow-up is required, he or she indicates what he or she plans to do as an intervention plan with this patient (Fig. 15).

Support Marker		Obstacle marker	
0	Beliefs and practices	0	Beliefs and practices
0	Hopes	0	Hopes
0	Relationships	0	Relationships
0	Values and commitments	0	Values and commitments
0	Transcendental experience *		

Fig. 14 Markers in support of the intervention/obstacle. (Reproduced with permission from Centre Spiritualitésanté de la Capitale-Nationale' (CSsanté). Copyright © 2019 CSsanté. All rights reserved)

---

[27] This approach differs from the one presented before, in that the patient is no longer in a stage of developing meaning but in a stage where, in the interventions, we reflect the main axes of meaning in his or her life.

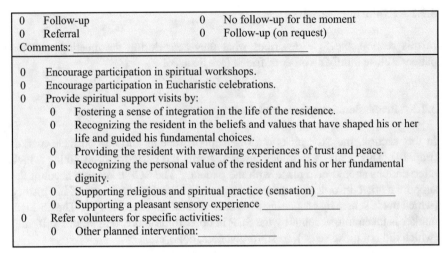

0	Follow-up	0	No follow-up for the moment
0	Referral	0	Follow-up (on request)
Comments: _____			

0	Encourage participation in spiritual workshops.
0	Encourage participation in Eucharistic celebrations.
0	Provide spiritual support visits by:
	0	Fostering a sense of integration in the life of the residence.
	0	Recognizing the resident in the beliefs and values that have shaped his or her life and guided his fundamental choices.
	0	Providing the resident with rewarding experiences of trust and peace.
	0	Recognizing the personal value of the resident and his or her fundamental dignity.
	0	Supporting religious and spiritual practice (sensation) _____
	0	Supporting a pleasant sensory experience _____
0	Refer volunteers for specific activities:
	0	Other planned intervention:_____

Fig. 15 Follow-ups and plan of intervention. (Reproduced with permission from Centre Spiritualitésanté de la Capitale-Nationale' (CSsanté). Copyright © 2019 CSsanté. All rights reserved)

3.3 In Addition... Sensation and the Transcendent Experience

Sometimes patients, their relatives or close friends might say something about their sensations (relating to the five senses). In some cases, it is difficult for the SCP to establish a link with one of the four key markers. However, it is very important, in our opinion, to be alert to these statements, which provide important information regarding sensation, which is one of the main dimensions of our transversal reference marker, meaning. Here are some examples of how this can be expressed in a patient's speech. These elements of sensation can be indicated in the 'values/commitments' or in 'experience of transcendence' markers. They are privileged forms of spiritual support.

Examples:

- Yesterday I watched a magnificent sunset from my room.
- I smelt the roses that my son-in-law brought me: they were wonderful.
- I still enjoy the taste of good food.
- This music soothes me.
- On my travels, the warm sand gave me a sense of freedom.

3.4 Impact on the SCP and Care Team

The RESS tool has allowed us to formalize the assessment process and serves as a basis for drawing up note-writing guides setting out clear steps in the process. The new *Quebec model of recording spiritual care* has several benefits for our profession; in particular, it helps to refine the clinical assessment process. Indeed, the inclusion of markers encourages the SCP to review the assessment meetings by asking a number of fairly open questions which provide relevant guidelines for making an assessment. We have observed that the SCP develops a better understanding of the patient's experience through this process and thus improves their reporting activities as a whole. Finally, our clinical workshops, which aim at reviewing our support methods, have benefited both from the RESS assessment tool and from this new way of writing the note to file.

The tool also fosters interprofessional dialogue. In fact, the well-defined structure and sequencing of the notes facilitates comprehension, speed of reading and the retention of information in the medium term. At the same time, the medical and nursing staff are better able to understand the service we provide, and information can be more easily passed on. Finally, we are convinced that a well-conceived process for assessing and writing notes to file promotes a more complete understanding of the patients, which can only serve their well-being and the quality of their stay in hospital.

4 Conclusion

In the sixteenth century, Ignace de Loyola recalled that one who determines little understands little and helps even less (De Loyola 1991, 647).

It is in this spirit that we have sought to improve, in recent years, our processes for preparing the note on file. In particular, we focused on the evaluation note, which is crucial in the development of the care plan. A lot of effort has been put into making it as accurate as possible so that it can fulfil its purpose: to help!

The systematic writing of the spiritual care note is still a very new practice in Quebec – it's barely 15 years old! Even today, it still presents many challenges, the most important of which is inherent to our profession. Indeed, spirituality remains a complex and extensive notion for which there is no definition shared by the entire scientific community. Hence the difficulty of developing a writing framework that is flexible enough to allow us to record our assessment in relation to our vision of spirituality and that is structured enough to be read by the interdisciplinary team in a user-friendly and supportive manner.

In the work of the spiritual care provider, the note on file constitutes a structuring practice that reflects the content of the interventions carried out but also the meaning of spiritual accompaniment that has emerged over the years. We believe that the practice of writing notes will continue to challenge our profession by regularly forcing us to rethink our 'theology' and what we convey as a vision of spirituality in a secular and demanding context in terms of the quality of accountability.

Appendices

Appendix A: Features of a Note to File

Both an assessment/intervention note and a development note must have certain features in order to perform the functions that are intended (Brassard 2000; CPSSS 2008).

Clarity

Clarity in a note can be achieved in a number of ways: it must be precise, concise, chronological and readable. In other words, the note to file tells the key elements of a spiritual care intervention in just a few words and can be quickly grasped and understood by other professionals and by the users themselves. 'A note is precise when it contains no superfluous elements and does not allow room for interpretation. It is understood in the same way by all those who read it and does not contain any uncertainty' (Brassard 2000, 13).

Readability

A note must be easily readable and free of spelling and syntax errors. These last two elements may seem unnecessary, but in fact they promote clear communication and reflect the professional quality of the services provided. In addition, they reduce the risk of a note being misinterpreted. In other words, a note which is clear is one that constitutes a coherent and easily understandable whole.

Reliability

It is assumed that what is written in the note to file is true and reliable, so that any user consulting their file would recognize themselves in what was written and would not be hurt by it. Anything recorded is truthful, that is, it reflects what has been said and done in reality.

Relevance

A relevant note is one which reports on the content of the interview and what was observed, heard and seen with the user. It describes what they said, their behaviour and their reactions in a specific context or situation. If, occasionally, the note refers to facts about the surroundings or reactions from other people, it should only do so if these shed light on the user's own experience and reactions.

As we will see later on, the assessment note may mention actions taken by the SCP. The note also describes the effect of the intervention on the user: how do they react to it, and what effect does it have on them?

Objectivity

An objective note allows us to understand unequivocally the SCP's professional opinion on what the patient expresses and their analysis and objectives.

Writing Time

Finally, the note to file should be written as quickly as possible; the information should be recorded when it is fresh in the mind and the memory is still reliable. If, exceptionally, a note to file cannot be written on the same day as the meeting with the user, the SCP must write the note later on. In this case, the SCP records the date and time at which the note is written and writes in the past tense, specifying in the note the date and time of the meeting with the user.

Source: Centre Spiritualitésanté de la Capitale-Nationale' (CSsanté). Reproduced with permission. Copyright © 2019 CSsanté. All rights reserved

Appendix B: Identification of a Support Marker/Obstacle Marker

Identification of a support marker/obstacle marker (beliefs/practices, hopes, relationships, values/commitments)		
Support marker		
Significance (ability to understand) Does the identified marker help the patient understand what is happening to him/her?	Direction (ability to project oneself) Does the identified marker help the patient to envisage the future?	Sensations (felt positive) Does the identified marker bring the patient good sensations or help him/her to feel something good?
Obstacle marker		
Significance (inability to understand) Does the identified marker generate questions or misunderstandings?	Direction (inability to project oneself) Does the identified marker generate fears or questions about what is coming?	Sensations (felt negative) Does the identified marker generate negative sensations or negative feelings?

Commentary

Ralph Kunz (✉)
University of Zurich, Zurich, Switzerland
e-mail: ra.kunz@bluewin.ch

In Germany and Switzerland, pastoral care in hospitals is mostly a matter for the church. The Quebec model of recording spiritual care has a different cultural and organizational background and can therefore seem strange to Swiss and German chaplains. First of all, whenever a professional intervention takes place, it has to be recorded In Quebec, admissible interventions are assigned a value in "units of measurement" and care providers are required to compile a record of all such interventions. This means: For each unit of measurement reported, there must be a note on file. Bruno Bélanger's report strikingly illustrates contextual embeddedness of the model, which has been in development since 2002. It is important to know the whole story, which Bélanger and others tell. There is a debate in Switzerland about the lessons to be learned from the Quebec model. Maybe the Canadian experience will help assuage some fears; maybe it will fuel new fears.

What Bélanger et al. call a paradigm change shows up in the nomenclature among other places. The professionals formerly known as chaplains have, since 2011, been called "spiritual care providers," and they are no longer mandated by their church. The shift from the pastoral role of an emissary of the church to a clinical embeddedness within the organization weakens the religious profile of the profession, but strengthens the position of the profession. The essential question is now how a note should be made. It is no longer questioned whether spiritual care should be charted. The question of how to make a note becomes essential. There is mention of comprehensible language as well as a rationale for the intervention taken, which has to appear reasonable to members of the other professions involved in the treatment. The debate is all about implementing and evaluating a suitable tool that serves others, but also takes into account the needs of one's own profession.

In this context two characteristics of the Quebec-model are exciting. What are called "Markers for Spiritual Care Assessment" (RESS) in Quebec are given in order to structure and format the charting. The common goal of the "treatment" provides the structure, while special attention to the spiritual realm demands a corresponding terminology Bélanger et al. make it clear that they think of this as essential. A degree of uniformity increases intelligibility and facilitates the provision of spiritual care. The diagram in the first conceptual part of the manual provides information about the anthropology which guides the work of spiritual care providers. In my opinion, the core of the anthropology is most inspiring and theologically coherent. Bélanger et al. call the center a "transcendent experience." In the chapter, the origin of this term is only briefly discussed. However, there are indications enough of the originality of the diagram. Thus, the mention of Meister Eckhart suggests traces of the "divine spark" and the "fruit of the spirit" marks an interest in biblico-theological reflection upon the spiritual moment in a biblico-theological fashion. This is remarkable and rather unusual. When talking about spirituality in the context of spiritual care, we usually move in a theology-free zone. What "sense" and "well-being" mean is left open from an ideological point of view. They can be specified in different ways. Of course, these dimensions also occur in the present diagram. But meaning is, characteristically, a "shell" around the spiritual core. The center is therefore an experience "which goes beyond normal life." To call it "transcendent" is to assert its unavailability.

Bélanger et al. consider this transcendent innermost core as a potential: it can be intuited, but it can't be manipulated; it eludes treatment and calls for expert guidance. What is the charm of this model? It allows the proprium of the work of the spiritual care provider to be communicated in an adequate way. Through the courageous centering of transcendent experience, it can also be made plausible that the spiritual process triggered by the crisis of a disease can lead to personal transformation. What spiritual care providers can do differs from what a doctor, a nurse, or a therapist can do. Spiritual care providers accompany patients through change, while testifying to that change and searching for a language to be found "in the nook and crannies of everyday life." The Note-Writing-Guide is geared to this central task! This is very impressive!

Nevertheless, I think the question can be raised whether there are theological pathways to determining the transcendence Christologically. It is certainly no coincidence that spiritual-theological terminology is based on pneumatology. But this does not mean that Christological explication would be superfluous. I think that theological self-reflection on the part of the Christian spiritual care provider could deepen and differentiate the transcendent experience.

To summarize, for me the Quebec Model is highly persuasive. Despite its different background to Swiss context, it is an attractive and, in many ways, stimulating example of a charting culture in a clinical setting. The spiritual care provider participates in a joint recording project. I like the idea of a container which contains all the relevant information necessary to give the patient the best treatment possible. However, the European reader will be struck by the fact that there is not even a rudimentary debate about the potential risks posed by the contribution of documentation by spiritual care providers. Among other things, this has to do with professional self-image. Spiritual care providers conceive themselves as part of a team that has a common goal for which the user's social and health care record is an essential tool: "One of its key functions is to promote communication between all those involved in the patient's treatment." Bélanger et al. make it very clear that they support a disciplined charting culture and that they take the involvement of the spiritual care providers for granted. In my view, it is interesting that in Quebec the change of system, which was started in 2002 with a directive from the ministry, has been implemented in such a highly constructive way. What was more or less a forced change has been taken as an opportunity to reposition the spiritual care provider within the health system.

References

Brassard, Yvon. 2000. *Apprendre à rédiger des notes d'observation*, 3e édition, volume 1–2, Longueuil: Loze-Dion.

Bryson, Ken. 2015. Guidelines for conducting a spiritual assessment. *Palliative and Supportive Care* 13: 91–98.

Caenepeel, Didier. 2017. Conférence prononcée au CSsanté à Québec (Février 2017).

CPSSS. 2008. *Guide de rédaction de la note au dossier*, Centre de pastorale de la santé et des services sociaux, document inédit.

De Loyola, Ignace. 1991. *Écrits* (Collection Christus no 76). Paris: Desclée de Brouwer, Montréal: Bellarmin.

De Lubac, Henri. 1990. *La lumière du Christ*. Tome 1: *Théologie de l'Histoire*, 115–121. Paris: Desclée De Brouwer.

Dumas, Marc. 2010. La spiritualité aujourd'hui: Entre un intensif de l'humain et un intensif de la foi. *Théologiques* 18/2: 199–211.

Fromaget, Michel. 2007. *Naître et mourir, anthropologie spirituelle et accompagnement des mourants*, François-Xavier de Guibert, Paris. Document inédit.

———. 2008a. *De l'esprit comme objet d'expérience/introduction à la phénoménologie de l'esprit*, conférence prononcée à Québec. Document inédit.

———. 2008b. *Conception spirituelle de l'homme et accompagnement des mourants*, conférence prononcée à Québec. Document inédit.

———. 2009. Anthropologie et soins de santé, les trois dimensions de la personne humaine. *Spiritualitésanté* 2009: 12–22.

Kellen, Jacqueline. 2015. *Le bréviaire du Colimaçon*. Paris: Desclée et de Brouwer.

Marin, Claire. 2013. La maladie/entre crise et critique. *Spiritualitésanté* 6 (3): 14–16.

Pargament, Kenneth I. 2007. *Spiritually integrated psychotherapy/understanding and addressing the sacred*. New York/London: The Guilford Press.

Rosselet, François. 2002. Prise en charge spirituelle des patients: la neutralité n'existe pas. *Revue Médicale de la Suisse Romande* 122: 175–178.

Roy, Louis. 2014. *L'expérience de transcendance. Phénoménologie et analyse critique*. Mediaspaul: Montréal.

Rumbold, Bruce. 2013. Spiritual assessment and health care chaplaincy. *Christian Bioethics* 19 (3): 251–269.

Ugeux, Bernard. 2001. *Retrouver la source intérieure*. Ivry-sur-Seine: De l'Atelier.

Waaijman, Kees. 2006a. What is spirituality? *Acta Theologica Supplementum* 8: 1–18.

———. 2006b. Conformity in christ. *Acta Theologoica Supplementum* 8: 41–53.

Zundel, Maurice. 2005. *Je est un autre*. Paris: Éditions Le Sarment. (publication originale 1971).

Charting and Documenting Spiritual Care in Health Services: Victoria, Australia

Christine Hennequin

Diverse models of chaplaincy and spiritual care have existed in health services in Australia for decades. This overview of charting will address the development in the state of Victoria from the perspective of Spiritual Health Association as a peak body for spiritual care in health services (Spiritual Health Association 2020). Some of these developments have been driven by national standards in charting and data collection and consequently have had some impact in the spiritual care sector nationally. Funding sources and the models of delivery have evolved and changed quite significantly since the 1950s to a more professional model, and this evolution has necessitated changes in accountability, including charting and documentation.

1 Background/Rationale

In Victoria, a state in south-eastern Australia, chaplaincy has moved from mainly following a Christian-delivered model established by churches in the 1950s and 1960s to current models of spiritual care responding to all faiths and beliefs. In the 1950s the Anglican Church employed its first Chaplain at the Royal Melbourne Hospital in Melbourne (Kenny 2003). John Moroney established professional procedures in his role. Catholic Chaplaincy was well established by the 1960s with a part-time Catholic Chaplain employed at the Royal Women's Hospital in 1961 for sacramental ministry and education and gatherings of mental health and hospital chaplains from several denominations occurred in the 1950s and 1960s (Kenny 2003).

Today, most major metropolitan public hospitals in Melbourne and some regional hospitals in Victoria employ a Spiritual Care Manager, Director or Coordinator with varying numbers of professional staff employed by the health service or by a faith

C. Hennequin (✉)
Spiritual Health Association, Collingwood, VIC, USA
e-mail: development@spiritualhealth.org.au

© The Author(s) 2020
S. Peng-Keller, D. Neuhold (eds.), *Charting Spiritual Care*,
https://doi.org/10.1007/978-3-030-47070-8_5

community. Some private hospitals, especially denominational hospitals, employ spiritual care staff. Some health services conduct a Clinical Education programme concurrently and use trained and supervised Clinical Pastoral Education students as part of their resources. A few metropolitan and many regional health services use volunteers with a limited scope of practice as part of the model of providing spiritual care.

2 The 1990s: Developing Pastoral Diagnoses

The Austin Hospital joined other Allied Health departments in 1992 to develop codes to document their work. Graeme D. Gibbons, together with pastoral care supervisors and colleagues at the Austin Repatriation Hospital, developed diagnoses and interventions which were used to document pastoral care provision in an Allied Health statistical package (Gibbons 1998). As a Clinical Pastoral Education student doing my first unit at the Austin in 1993, I used these codes and participated in education regarding the diagnoses and interventions, their descriptions and interpretation as we reflected on our work on the wards. The Pastoral Care Casemix Project initiated by Gibbons and other medical centre chaplains was influenced significantly by Clinical Pastoral Education (Gibbons 1998) and was important in establishing pastoral care codes which "Integrated information technology with a foundational theological perspective and the traditions associated with CPE and hospital ministry" (Kenny 2003).

Incorporating these diagnoses alongside other allied health disciplines in the health services' database was an important step in demonstrating the contribution and value of pastoral care within the broader healthcare context. Research and analysis of the data from January 1993 to May 1998 provided information about the chaplains' core activities which included "promoting spiritual transcending" (57% of all entries), "promoting spiritual intactness" (36.1%) and "enacting ministry (6.2%)" (Kenny 2003).

Skills and interest in charting have varied and have relied for a long time on the personal interest or skill of spiritual care managers. In his thesis on chaplains in South Australia, C.C. Aiken found that some paid close attention to this aspect of their role, while others saw administration, including collecting statistics, as a frustrating aspect of their role as hospital chaplains (Aiken 2010). Yet both the relational and administrative aspects of the role are important.

3 Guidelines for Pastoral Care in Australia

In the early 2000s, health services in Victoria still had diverse and individual ways of documenting spiritual care. Work was being done on several fronts to set standards for assessing and recording pastoral interventions. In 2002, the Australian

College of Chaplains requested the inclusion of pastoral care intervention codes for the Third Edition of ICD10-AM (National Centre for Classification in Health 2002).[1] The pastoral codes were disseminated and enthusiastically received by members at the Australian Health and Welfare Chaplains Association's (AHWCA)[2] national conference that year. The AHWCA published the Health Care Chaplaincy Guidelines in 2004 after extensive consultation, review and evaluation of a draft document in 2002. These guidelines were based on "the UK 'Health Care Chaplaincy Standards' as a national response to formulating a systematic approach for pastoral care and chaplaincy services across Australia" (Australian Health and Welfare Chaplains Association Inc. 2004).

The Guidelines Map clearly states the importance of charting and accurate documentation of pastoral care services as one of the competencies necessary in a systematic approach to providing services:

- D2 Provide and record pastoral care services
- D2.3 Assess patient's pastoral diagnoses
- D2.4 Assess individual spirituality and strength
- D2.5 Chart pastoral interventions into patient records (where permitted)
- D2.6 Codify and enter details into pastoral care database
- D2.7 Retrieve data from database

(Australian Health and Welfare Chaplains Association Inc. 2004)

The Healthcare Chaplaincy Council[3] of Victoria commissioned further work to produce standards and a minimum dataset for reporting pastoral care in a consistent and accountable manner in public and private hospitals over the next decade (Healthcare Chaplaincy Council of Victoria Inc. 2012). Standards were also developed for the mental health and aged care sectors. Standardisation was required to encourage practitioners to describe and collect information using consistent language and categories and to improve data collection within their health services so that pastoral care interventions would be visible and reported to management.

---

[1] ICD-10-AM is the International Statistical Classification of Diseases and Related Health Problems, Tenth Revision, Australian Modification. It consists of a tabular list of diseases and accompanying index.

[2] AHWCA, the Australian Health and Welfare Chaplains Association, was the association representing chaplains until 2010. It was replaced by the new national association Spiritual Care Australia in February 2010.

[3] The Healthcare Chaplaincy Council of Victoria and Spiritual Health Victoria are the precursors to Spiritual Health Association, the peak body for spiritual care in health services. The name changes occurred in June 2014 and July 2019, respectively.

4   A New Minimum Dataset

In 2013–2014 the Department of Health and Human Services requested that Spiritual Health Association report on spiritual care activity from public hospitals as part of its funding requirement. The department was particularly interested in information about spiritual care interventions by clergy and faith representatives as well as faith-based volunteers. Part of Spiritual Health Association's funding from the department is disbursed to faith groups, and increased accountability was required of them. Spiritual Health Association kept abreast of international developments in the area of documentation and reporting and followed with interest the work that was being done by Dean V. Marek in Milwaukee regarding identifying the cost of chaplaincy interventions (Marek 2005) and of Massey et al. regarding a taxonomy of terms for spiritual care (Massey 2015). The requirement for effective documentation and clear accountability and the focus on quality and safety remains following the recommendations of Targeting Zero: the review of safety and quality assurance in Victoria (Safer Care Victoria 2018; Spiritual Care Australia 2018).

The pilot project engaged spiritual care managers from 23 health services who collaborated with Spiritual Health Association to provide the data for spiritual care activity. During the process of data collection, it became clear that they were not collecting data in a consistent manner. There were no specific guidelines for the sector, and some practitioners were unaware of existing national standards or did not know how they were to be applied in data collection. A Working Group consisting of nine spiritual care managers from metropolitan hospitals met with Spiritual Health Association staff over eighteen months to develop a new Spiritual Care Minimum Dataset Framework based on current health data standards and definitions (Spiritual Health Victoria 2015). Once completed, the framework was circulated to Chief Executive Officers of health services with a spiritual care department and to spiritual care management at those sites. Education sessions were held by Spiritual Health Association for managers and senior practitioners, and a workshop was presented at the national Spiritual Care Conference in Sydney in 2016 with an additional framework: Spiritual Care in Victorian Health Services: Towards Best Practice Framework. This framework helped to improve the quality and consistency of data collection as part of best practice. There has been national interest in the framework as it provided guidance in the use of the ICD-10-AM/ACHI/ACS[4] codes

---

[4]ACHI, the Australian Classification of Health Interventions, has also been in use since 1998. ACHI was based on the Medicare Benefits Schedule (MBS) and was previously known as the Medicare Benefits Schedule-Extended (MBS-E). The National Centre for Classification in Health developed it with assistance from specialist clinicians and clinical coders. ACHI codes have seven digits. The first five digits are the Medicare Benefit Schedule (MBS) item number where one exists. The two-digit extension represents specific interventions included in that item. The classification is structured by body system, site and intervention type. Interventions not currently listed in MBS have also been included (e.g. dental, allied health interventions, cosmetic surgery). ACHI consists of a tabular list of interventions and accompanying alphabetic index. ACS, the Australian Coding Standards, have been developed with the objective of satisfying sound coding conventions

which are used nationally (Independent Hospital Pricing Authority 2018) to report on activity-based funding. The framework has also been used to establish a minimum dataset for a new spiritual care service at the Central Adelaide Local Health Network in South Australia (Bossie 2018).

5   Towards a Best-Practice Approach

The focus on documentation and data collection and the significant work undertaken by the Healthcare Chaplaincy Council of Victoria and subsequently by Spiritual Health Association over the last two decades has assisted our sector in Victoria to become more competent in this area. Preliminary data from a recent state-wide survey conducted by Spiritual Health Association in 2019 suggests that there has been an increase in the number of spiritual care departments using the ICD-10-AM/ ACHI/ACS intervention codes (Spiritual Health Association 2019b). Documenting in medical records is an essential requirement in a complex health service environment. Spiritual Care departments have worked to improve this aspect of reporting in Victoria. While the skills of practitioners vary, there are some managers who are very interested in and engaged with data collection and charting. Ongoing consultation with spiritual care managers and practitioners through regular updates, workshops and Spiritual Care Management Network meetings keeps data collection and charting on the agenda (Spiritual Health Victoria 2018; Spiritual Care Australia 2018).

As part of its goal and mission,[5] Spiritual Health Association has identified and responded to the strategic issues for spiritual care within healthcare and encouraged the sector in Victoria to support and collaborate with it. Spiritual Health Association's Capability Framework 2016 (Spiritual Health Victoria 2016) and the Spiritual Care Australia Standards of Practice (Spiritual Care Australia 2014) recognise documentation as an essential skill for practitioners and for volunteers. Developing this as a foundational skill when delivering Clinical Pastoral Education and when training new practitioners is paramount for the future.

The Spiritual Care Minimum Data Set Framework was evaluated in late 2016 by surveying spiritual care management and was reviewed in 2018. Evaluation results were positive and demonstrated that this significant work has assisted our sector in Victoria to become more engaged with data collection and to recognise its importance in the provision of healthcare. Ongoing consultation with managers, and practitioners, health service executives and health information administrators during the revision has ensured that the new guidelines reflect current practice, standards and language.

---

for use with ICD-10-AM and ACHI. They apply to all public and private hospitals in Australia. The ongoing revision of the Australian Coding Standards ensures that they reflect changes in clinical practice, clinical classification amendments and various user requirements of inpatient data collections.

[5] Spiritual Health Association's mission is to enable the provision of quality spiritual care as an integral part of all health services in Australia.

The new guideline "Spiritual Care in Medical Records: A Guide to Reporting and Documenting Spiritual Care in Health Services" (Spiritual Health Association 2019a) aligns with current practice in health services and the current National Safety and Quality Health Service Standards (Australian Commission on Safety and Quality in Health Care 2017) as well as the updated Spiritual Intervention Codes from the Australian Consortium for Classification Development (Independent Pricking Authority 2018). The guideline was circulated to Victorian health services and spiritual care managers and to other key stakeholders in February 2019. It is available on the Spiritual Health Association website (Spiritual Health Association 2019).

Spiritual care departments that are well integrated adhere to their health service's policy and to the current guidelines on documenting in medical records. As a result, communication within the multidisciplinary team is enabled and standards are met. In addition, the guidelines assist in educating spiritual care staff and students about documentation and charting. Staff can develop their capability in that area and meet the requirements of their role.

The following are case studies of documentation in three health services in Victoria, Australia. Data collected electronically and in paper-based form by the health service is used in different ways for reporting at different levels of the health system:

- By the spiritual care department to collect a detailed account of spiritual care provision by identifying patients' needs, spiritual care interventions and outcomes
- By the spiritual care department to report to management for internal reporting of the number of episodes of patient care including direct patient contact and indirect contacts such as administration and travel
- By the health service to report to the Department of Health and Human Services in Victoria by mapping and transmitting aggregate data according to the reporting requirements and business rules each financial year[6]

6   Case Studies

6.1   *The Royal Melbourne Hospital: Parkville, Victoria, Australia*

The Royal Melbourne Hospital is a large metropolitan hospital in inner Melbourne and is part of Melbourne Health. The hospital still uses paper files in conjunction with an electronic system, Patient Flow Manager, which manages the patient jour-

---

[6] While spiritual care is not costed as part of Casemix funding in Victoria (Victoria State Government 2018a, b, c), it is included with other Allied Health interventions as part of the Not Automatically Qualified for Admission List (NAQAL) in the Victorian Admitted Episodes of Care Dataset procedure code list 2017–2018 (Victorian state Government 2018).

ney from admission to discharge. It includes a referral system where Allied Health referrals, including spiritual care, can be made. Currently, some medical records are still paper-based.

Multidisciplinary colleagues communicate in the Patient Flow Manager to request a referral from another discipline. Spiritual Care will accept the referral and provide an update to the referring colleague by providing information about the spiritual care interventions and time frames in which follow-up will occur.

Figure 1 is an image of the Patient Flow Manager system.

In addition, spiritual care practitioners use free text when writing Clinical Progress Notes (Fig. 2) by hand in medical records. The hospital will eventually have all existing records digitalised as part of the electronic medical record system. Practitioners are urged to document significant information in the patient's Clinical Progress Notes.

Verbal communication is encouraged, especially for significant care issues (i.e. organ transplant, family meetings) and at times for debriefing between colleagues. Melbourne Health's MH05 Documentation and Records Management Policy states this clearly under Procedure 5.4b.

Documentation at the Royal Melbourne Hospital meets the recommendations of Spiritual Health Association's guidelines for documentation (Spiritual Health Victoria 2019).

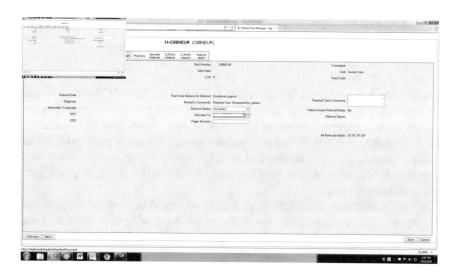

Fig. 1 Patient Flow Manager, Royal Melbourne Hospital. (Reproduced with permission from Melbourne Health. Copyright © 2020 Melbourne Health. All rights reserved)

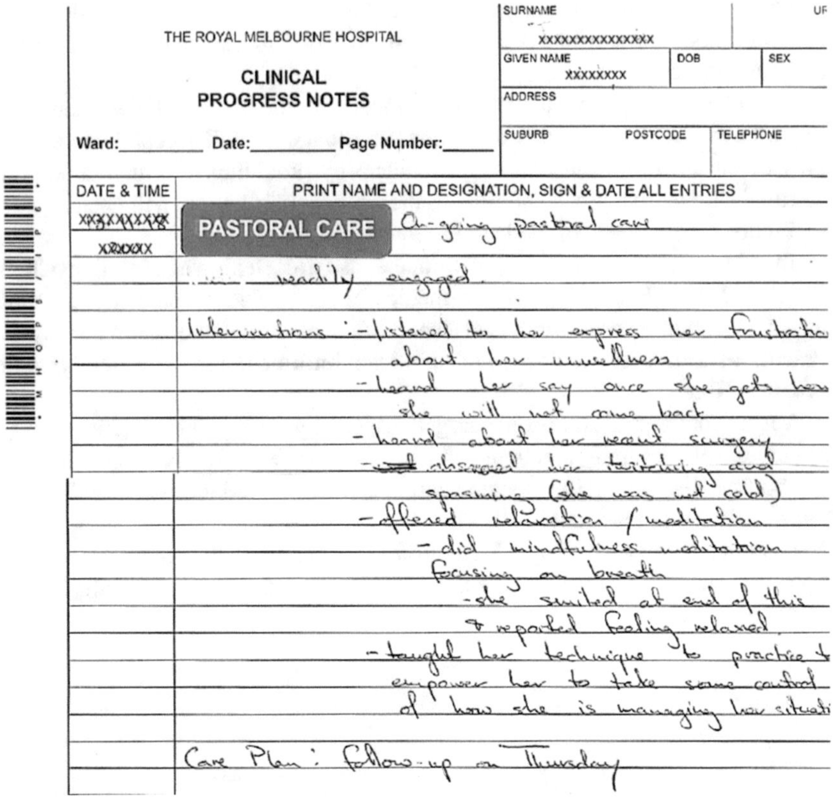

Fig. 2 Example of documentation, Royal Melbourne Hospital. (Reproduced with permission from Melbourne Health. Copyright © 2020 Melbourne Health. All rights reserved)

6.2   Bendigo Health

Bendigo Health is a regional health service in the state of Victoria. Bendigo Health employs a Manager of Pastoral Care and a Chaplain, both part-time. In addition, the Pastoral Care Department relies on a number of faith community representatives and volunteers to provide spiritual care within a specified scope of practice (Spiritual Health Victoria 2016). Documentation takes several forms: electronic and paper-based. Volunteers do not have access to the medical records, in accordance with health service policy.

To enable chaplains and religious visitors to document their visits, the Manager of Pastoral Care requested the development of form MR94H "Pastoral Care Intervention" (Fig. 3) to facilitate the documentation of four ICD-10-AM spiritual care interventions. The form could then be inserted in the medical record by Bendigo Health staff and later scanned to form part of the Digital Medical Record.

Those identified as volunteers can "tick the box" to indicate which intervention they have provided without having access to the Digital Medical Record, which is

FBH 335 750

BENDIGO HEALTH

PASTORAL CARE INTERVENTION

UR NO: .... MH NO:..

SURNAME:.. GIVEN NAMES:..

D.O.B:.. SEX:..

ADMISSION DATE: ....

CONSULTANT: .... WARD/CLINIC:..

USE LABEL IF AVAILABLE

This section must be filled in by all Chaplains and Religious Visitors

Name _____   Date _____

Signature _____   Time _____

Role: ☐ BH Employed Chaplain   ☐ Accredited Chaplain   ☐ Religious Visitor

Duration of Contact (in minutes) ☐ < 5   ☐ 5-10   ☐ 10-20   ☐ 20-40   ☐ 40 – 60   ☐ > 60
(Only use <5 after referral if pt not available / accepting care)
Ward
Service
Other

Intervention: Please tick one

Pastoral Assessment
☐ Formal Pastoral Care assessment

Pastoral Ministry or support
☐ Introducing the pastoral service
☐ Providing a listening presence
☐ Providing emotional support
☐ Providing grief/loss/transition/dying support
☐ Being present in times of uncertainty/crisis

Pastoral Counselling or Education
☐ Providing guidance on matters involving spiritual/religious meaning, belief or practice and ethical concerns

Pastoral Ritual/Worship
☐ Informal prayer
☐ Formal prayer/ritual/meditation/reflection
☐ Patient not accepting Pastoral Care
☐ Person not available (attempted visit after referral, ongoing relationship, listed as specific faith etc.)

This section must be filled in by BH employed Chaplains and may be filled in by other chaplains / religious visitors

Notes: (Note if intervention also includes family members, birthday visits, specific religious rites eg. anointing of the sick, or is for a family member *instead of* patient)

Religion
The patient/resident has consented to update patient record from _____ to _____
A request needs to be made to the ward clerk or Manager of Pastoral Care to action this update request.

Referrals

Incoming Date of Referral _____
From (name if known) _____
☐ Allied Health   ☐ Chaplain
☐ Doctor   ☐ Family member/friend
☐ Nurse   ☐ Religious community
☐ Self referred   ☐ Volunteer
☐ Ward Clerk   ☐ Other_____
Reason
☐ Listening presence   ☐ Emotional support
☐ Grief/Loss/Transition/Dying ☐ Uncertainty/Crisis
☐ Spiritual/Religious/Ethical guidance
☐ Prayer/Ritual/Meditation/Reflection
☐ Introduction of service
☐ Formal Assessment   ☐ Other_____
Source
☐ Phone ☐ Email ☐ Verbal ☐ Patient Flow Manager
Awareness/Consent
☐ Patient was aware of incoming referral
☐ Patient was NOT aware of incoming referral

Outgoing Date of Referral_____
An outgoing referral ☐ has been OR ☐ will be made to:
☐ Allied Health e.g. Social Work
☐ BH Employed Chaplain   ☐ Accredited Chaplain
☐ Religious community/Religious Visitor
☐ Other_____
Reason
☐ Listening presence   ☐ Emotional support
☐ Grief/Loss/Transition/Dying ☐ Uncertainty/Crisis
☐ Spiritual/Religious/Ethical guidance
☐ Prayer/Ritual/Meditation/Reflection
☐ Formal Assessment   ☐ Other_____
Method
☐ Phone ☐ Email ☐ Verbal ☐ Patient Flow Manager
Awareness/Consent
☐ Patient is aware of outgoing referral
☐ Patient is NOT aware of outgoing referral

☐ Includes family/friends (e.g. Bereavement support)

May 2016

PASTORAL CARE INTERVENTION

MR94H

Fig. 3 Example of documentation, Bendigo Health. (Reproduced with permission from Bendigo Health. Copyright © 2020 Bendigo Health. All rights reserved)

Fig. 4 Bendigo Health example. (Reproduced with permission from Bendigo Health. Copyright © 2020 Bendigo Health. All rights reserved)

outside of their scope. Volunteers can provide further notes about the visit and details regarding referrals as appropriate. The form avoids the duplication of requests for spiritual care and enables effective communication with the rest of the caring team.

Spiritual care practitioners and chaplains employed by their faith communities also use the MR94H form and have full access to the Digital Medical Record. Their notes would usually be more extensive to describe the spiritual care intervention provided.

The use of a digital referral flow chart has improved the efficiency of referrals. A SharePoint site is currently being developed for the Pastoral Care Department (Fig. 4). It will provide a record of completed Clinical Progress Notes MH94H forms, and chaplains will be able to access a history of spiritual care interventions for individual patients.

The Pastoral Care Department at Bendigo Health has devised a clever way to document spiritual care interventions in their medical records, while aligning itself with the health service's policy regarding who is authorised to access them. The department meets Spiritual Health Association's minimum dataset requirements. Its use of the four spiritual care interventions and descriptions of the pastoral encounter provide a common language for chaplains and volunteers to describe the spiritual care provided.

6.3   St. Vincent's Public Hospital: Fitzroy, Victoria, Australia

St. Vincent's Hospital is a large public hospital in inner Melbourne. Currently, St. Vincent's Hospital uses a combination of handwritten Clinical Progress Notes in paper-based medical records and electronic medical records systems for documentation. The Clinical Progress Notes are used to document the spiritual care intervention and to record the relevant themes which have emerged as part of the spiritual care encounter with the patient. The Clinical Progress Notes help to communicate with the multidisciplinary staff and provide an opportunity to educate staff about the contribution of spiritual care to patient care. A typical clinical note recording a Pastoral Encounter would read as follows:

PASTORAL CARE: 29/11/18 10.00am. Referral received with thanks to provide pastoral care for *<insert first name>* given teary on admission. Initial visit with *<insert first name>* to introduce pastoral care. Compassionate presence and active listening provided, enabling *<insert first name>* to share some life stories and reflections with themes of grief and loss. Pastoral support provided. *<insert first name>* expressed appreciation for the visit and requested a follow up visit which I will provide.

In one ward, clinical notes are recorded in InfoMedix – an electronic medical records system which allows for free text (Fig. 5).

In addition, spiritual care practitioners use the Patient Administration System (PAS) to record daily statistics of patient encounters. Practitioners record whether the encounter was as a result of a referral or self-referral (Comment 1 in Fig. 6) and the time taken. Comment 2 records the outcome of the encounter with an agreed descriptor as per the "Outcome Code" list. This provides common descriptors and consistent language for the practitioners to record their visits.

Descriptions of the four ICD-10-AM/ACHI/ACS spiritual care interventions are included in the drop-down boxes for "Service Type". Rituals such as "anointing" or "communion" are identified specifically as these are an important part of service provision in a Catholic public hospital.

Multidisciplinary feedback is positive regarding the contribution of spiritual care to patient care. The Pastoral Services Department are currently developing internal guidelines for patient documentation. St. Vincent's Hospital Melbourne meets Spiritual Health Association's guidelines for patient documentation (Spiritual Health Victoria 2019).

7 Conclusion

Advocacy for the ongoing integration of spiritual care within the health service is a priority for Spiritual Health Association. Outcomes of integration enable continuous quality improvement and meet healthcare standards as well as standards from Spiritual Care Australia. Spiritual care practitioners documenting in medical records contribute to this integration and to accountability.

Current research in Australia will add to the development of evidence-based outcome measures for spiritual care. Spiritual Health Association is collaborating with La Trobe University, Melbourne, and five health services to investigate the expectations for and the benefits from spiritual care provision. Further work on phase two and three will be undertaken this year (Spiritual Health Association 2019).

As a peak body, Spiritual Health Association needs to demonstrate that funding provided by the state government improves the quality of spiritual care in health services by ensuring evidence-based best-practice spiritual care.

The Victorian Department of Health and Human Services in 2016 has included spiritual care as an Allied Health profession in Victoria (Spiritual Health Victoria 2016). Spiritual Health Association and its representatives are invited regularly to represent the spiritual care sector on various Allied Health committees and forums

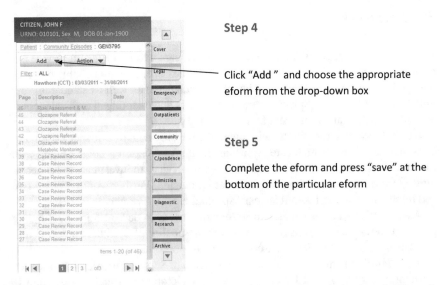

Adding an eform

Step 1

Open the record of the appropriate patient in the usual manner

Step 2

Select Tab for appropriate eform

In this case I have chosen Community

Step 3

Choose the Episode to which you wish to add the eform

In this case I have chosen Hawthorn CCT

Step 4

Click "Add " and choose the appropriate eform from the drop-down box

Step 5

Complete the eform and press "save" at the bottom of the particular eform

Fig. 5 InfoMedix, St. Vincent's Public Hospital. (Reproduced with permission from St. Vincent's Hospital Melbourne Limited. Copyright © 2020 St. Vincent's Hospital Melbourne Limited. All rights reserved)

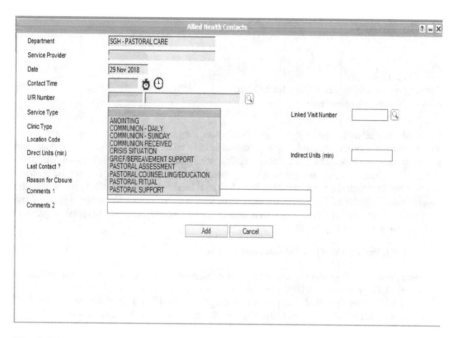

Fig. 6 PAS, St. Vincent's Public Hospital. (Reproduced with permission from St. Vincent's Hospital Melbourne Limited. Copyright © 2020 St. Vincent's Hospital Melbourne Limited. All rights reserved)

in the Department of Health and Human Services. The full integration of spiritual care into the health system is yet to be achieved, and there are still no mandatory standards for documentation (Holmes 2018). However, having spiritual care represented at state Allied Health forums ensures that we contribute to Allied Health National Best Practice Data Sets (Victoria State Government Health and Human Services 2017) and continue to improve the quality of data that we, as a sector, provide within the health system.

Commentary

Livia Wey-Meier (✉)
Theological Faculty of Chur, Chur, Switzerland
e-mail: livia.wey@mei-wey.ch

Christine Hennequin's contribution speaks a different language than the previous contributions. The question of documentation is not approached here as a further reflection on the conversation had in the hospital room and on its confidentiality, nor is the internal struggle for standardisation in recording pastoral encounters put in the foreground. The article unfolds rather from the perspective of a health-political administrative office – Spiritual Health Association (SHA) is the umbrella organisation for Spiritual Care in health services in Victoria – the complete history of pastoral and spiritual services in this region.[7] Three examples of a given minimum dataset to report on pastoral and spiritual services are provided.

1   From Pastoral to Spiritual Intervention Codes

The fact that this minimum dataset uses an official, globally accessible coding system that is almost 20 years old to record pastoral interventions may surprise some readers. In July 2002, four major "pastoral intervention" codings ("WHO-PICs") were incorporated and made available as part of the ICD-10-AM, which can be used by chaplains (or other pastoral and spiritual care workers) to record their interventions with patients and other clients (e.g. family and staff). These four major interventions are called (1) pastoral assessment, (2) pastoral ritual and worship, (3) pastoral ministry, and (4) pastoral counselling or education. Originally developed in Australia, these codes have hardly gained importance outside of Australia and have aroused little interest. Accordingly, they have rarely been evaluated in research or practice. One name from the Australian university research community that must be mentioned in this context is Lindsay B. Carey of La Trobe University. He himself has led some research projects on the codings and found that hardly any research had been done on them. The codes of interventions are unknown in many countries, which has led to other "attempting to reinvent the coding wheel". As regards the content of the codes, Carey makes various proposals for revision in an article from 2015. Among other things, he suggests moving away from the Christocentric formulations of the codes. He advocates supplementing "pastoral" with "religious" and "spiritual". The abbreviation PIC for "Pastoral Intervention Codes" therefore becomes REPSIC for "Religious, pastoral, spiritual Intervention Codes".[8]

The extent to which Spiritual Health Association is familiar with or has taken up Carey's research is not discussed in Hennequin's article. But attentive readers will have noticed that Hennequin's contribution indicates a further development of the intervention codes: while,

---

[7] Spiritual Health Association (SHA) as a "state peak body" for advancing the quality and availability of spiritual care in health services is supported by the State Government of Victoria and advised by a multi-faith board and council. SHA works in collaboration with all the faith communities that are SHV-members: Anglican, Baptist, Buddhist, Roman Catholic, Church of Christ, Hindu, Islamic Council, Jewish Council, Lutheran, Macedonian, Orthodox Christian Chaplaincy Council of Victoria, Presbyterian, Salvation Army, Sikh and Uniting Church. The following reflections therefore concern all the SHA-members and their spiritual care providers equally. For further information, see http://www.spiritualhealth.org.au/about.

[8] For the whole section, see Carey, L.B., and J. Cohen. 2015. The Utility of the WHO ICD-10-AM Pastoral Intervention Codings within Religious, Pastoral and Spiritual Care Research. *Journal of Religion and Health* 54:1772. 10.1007/s10943-014-9938-8

in section three of the article, she speaks of Pastoral Codes, she refers to "updated Spiritual Intervention Codes" in section five. These are the official intervention codes currently in force, approved by the Australian Consortium for Classification Development in 2017: (1) Spiritual assessment; (2) Spiritual counselling, guidance or education; (3) Spiritual support; (4) Spiritual ritual; and (5) Allied health intervention, spiritual care. These updated codes incorporate many of Carey's recommendations, but they do not use the terms "religious" and "pastoral".

The intervention codes form the core of SHA's provision the reporting and documenting of spiritual care. The data that can be drawn from their use might thus be described as the minimum dataset to be collected according to SHA. The coding itself, however, does not yet say anything about the method to be used for documentation. The codes are not a method, not a tool, but a framework. When asked in the discussion, Hennequin pointed out that the given coding of interventions of the different spiritual care givers is no substitute for the careful choice of methods for the documentation of Spiritual Care. Rather, SHA only elaborates the general guidelines in which it specifies how something should be done. This also has to do with the fact that the different hospitals often use completely different information systems. While the interventions are classified (intervention codes) and must be adopted as specified by SHA; everything else is being adapted for each single system.

2   Structural Distance and Participation

Due to its organisational form, which is not operational but for the most part administrative, SHA needs reliable partners to implement its requirements on site. For this reason, Hennequin often points out in oral exchanges how important the maintenance of good relationships is for SHA's work. Ultimately, it is the functioning relationships with the local spiritual care managers in the hospitals that decide whether the guidelines drawn up for documentation address the questions and concerns of the practitioners and are therefore subsequently adhered to across the board. Unfortunately, the reader of the article does not learn anything about the evaluation of the content of the information sheets and documentation guidelines mentioned in section five. Especially in view of Australia's many years of experience with the intervention codes, these evaluations would be of great interest to other countries that are just beginning to standardise the documentation of spiritual care.

And what about the patients for whose sake a meaningful documentation is ultimately demanded and desired? The patients seem to be included in Hennequin's the previous article, but not mentioned. The change from Christian pastoral care to pluriconfessional spiritual care goes hand in hand with a new culture, as Hennequin emphasises, one which is oriented towards professionalisation and accountability. Documentation, in this context, stands in connection with the word "accountability". The intervention codes closely adhere to existing forms of medical documentation and suggest – albeit not in a billable form – an ordered, uniform and comprehensible procedure in the field of spiritual care, a professionalised and accountable practice of spiritual care.[9] But is the application of the codes also perceived as appropriate? The voices of spiritual care employees are not the only relevant ones here. It would be extremely desirable and of great interest to gather opinions and

---

[9] Even though the interventions classified by the codes are not billable, the requirement for accountability has financial consequences for Spiritual Care Providers in Victoria. The Victorian State Government provides funding to SHA for the development and provision of spiritual care services in Victorian health services. Based on the 2016 census figures, half of the money is distributed by SHA to the faith communities that provide spiritual care in Victoria. In return, the communities are obliged to be accountable for their work. They present annual reports on financial expenditures, explaining what the communities do, where they work, whom they visit. If a report is missing, SHA can reduce the amount that the respective faith community receives annually to enable their provision of spiritual care services.

experiences from patients. The desideratum of patient participation also came up several times in discussion at the workshop. Awareness of this seems to exist, but implementation is still in its infancy – even though a research partnership between SHA and La Trobe University is under consideration. It would be interesting to know the results.

References

Aiken, C. 2010. *How we do Chaplaincy: A case study of South Australian Chaplains' understanding about their way of doing Chaplaincy.* Retrieved May 28, 2018, from https://repository. divinity.edu.au/517.

Australian Commission on Safety and Quality in Health Care. 2017. *National safety and quality health service standards,* 2nd Ed. Sydney, NSW, Australia. Retrieved March 18, 2019, from https://www.safetyandquality.gov.au/our-work/assessment-to-the-nsqhs-standards.

Australian Health and Welfare Chaplains Association Inc. 2004. *Health Care Chaplaincy guidelines.* Australia.

Bossie, C. 2018. *Data collection – Spiritual interventions codes from Jan 1st, 2018.* Adelaide, South Australia, Australia. Retrieved May 29, 2018.

Gibbons, G.D. 1998. Developing codes for pastoral diagnoses and pastoral responses in hospital chaplaincy. *Chaplaincy Today* 14 (1): 4–13.

Healthcare Chaplaincy Council of Victoria Inc. 2012. *HCCVI standards for reporting on pastoral care services in Victorian hospitals.* Abbotsford.

Holmes, C. 2018. Stakeholder views on the role of spiritual care in Australian hospitals: An exploratory study. *Health Policy.* Retrieved May 27, 2018, from www.healthpolicyjrnl.com/article/S0168-8510(18)30046-0/fulltext.

Independent Hospital Pricing Authority. 2018. *Classifications.* Retrieved from IHPA: www.ihpa. gov.au/what-we-do/non-admitted-care.

Kenny, J.M. 2003. *A finger pointing to the moon.* Mulgrave: John Garratt Publishing.

Marek, D.V. 2005. *How much does it cost for Chaplains' services?* Retrieved from National Association of Catholic Chaplains: www.nacc.org/vision/most-requested/how-much-does-it-cost-for-chaplain-services.

Massey, K. 2015. *What do I do? Developing a taxonomy of chaplaincy activities and interventions for spiritual care in intensive care unit palliative care.* Retrieved May 25, 2018, from BMC Palliative Care: bmcpalliatcare.biomedcentral.com/articles/10.1186/s12904-015-0008-0.

National Centre for Classification in Health. 2002. *ICD10-AM third edition education.* Retrieved from ACPE Research: www.acperesearch.net/ICD_10_AM_3rd_ed_Education.pdf.

Safer Care Victoria. 2018. *Home.* Retrieved from health.vic: www2.health.vic.gov.au/hospitals-and-health-services/safer-care-victoria.

Spiritual Care Australia. 2018. *Towards New Horizons: Spiritual Care Australia conference.* Program Guide, 14. Newcastle, NSW, Australia. Retrieved from Spiritual Health Victoria.

Spiritual Health Association. 2020. *About us.* Retrieved 27 February, 2020. https://spiritualhealth. org.au/about.

———. 2019. February. *Spiritual care in medical records: A guide to reporting and documenting spiritual care in health services.* Abbotsford, Victoria, Australia. Retrieved February 27, 2020, from https://spiritualhealth.org.au/standards.

———. 2019a. *Annual report 2018–2019.* Retrieved from https://spiritualhealth.org.au/about Annual Report-2018-2019.pdf.

———. 2019b. *The spiritual Care state-wide survey report 2019 – Preliminary report.* Victoria, Australia.

Spiritual Health Victoria. 2015. *Spiritual care minimum dataset framework.* Abbotsford, Victoria, Australia.

———. 2016. Standards and frameworks – Capability framework for spiritual care practitioners in Health Services 2016. Retrieved February 27, 2020, from Spiritual Health Association. https://spiritualhealth.org.au/standards.

———. 2018. *Spiritual Care Management Network meeting*. Melbourne, Victoria, Australia.

Victoria State Government. 2018a. *Hospitals and health services: Casemix funding*. Retrieved from health.vic: www2.health.vic.gov.au/hospitals-and-health-services/funding-performance-accountability/activity-based-funding/casemix-funding.

———. 2018b. *Medical records*. Retrieved from Freedom of Information: http://www.foi.vic.gov.au/home/foi/what+you+cannot+access/what+you+cannot+access+-+medical+records.

———. 2018c. *Allied health professions*. Retrieved from health.vic: www2.health.vic.gov.au/health-workforce/allied-health-workforce/allied-health-professions.

Victoria State Government Health and Human Services. 2017. National allied health standardised data development – Victorian Consultation. National. Melbourne, Victoria, Australia.

Victorian state Government. 2018. *VAED criteria for reporting procedure code lists 2017–18*. Retrieved from health.vic: www2.health.vic.gov.au/about/publications/policiesandguidelines/vaed-criteria-for-reporting-procedure-code-lists-2017-18.

Spiritual Care Charting/Documenting/ Recording/Assessment: A Perspective from the United Kingdom

Linda Ross and Wilfred McSherry

We were asked if we would provide a United Kingdom (UK) perspective on charting or documenting of spiritual care. This has been no straightforward task. We identified some of the dilemmas and challenges in assessing and documenting back in 2002 (McSherry and Ross 2002) and 2010 (McSherry and Ross 2010). We were hopeful that things may have changed; however, instead we found the situation has become even more complex. This is largely because of the number of people with responsibility for spiritual care, lack of clarity surrounding their roles and responsibilities in providing spiritual care and in documenting what they do, and conflicting terminology. Added to this the landscape of healthcare policy and practice is constantly changing meaning that healthcare staff have to be incredibly adaptable as their roles and responsibilities respond to these changes. An increasing awareness of Information Governance and concerns about protecting patient confidentiality means that staff (particularly chaplains) are cagy about what they are willing to share publicly in relation to their practice. In this chapter, we explore this complex situation in more detail. We start by giving a brief overview of the place of spirituality in the UK National Health Service (NHS). We then look at who provides spiritual care, identifying the challenges they face in so doing. The additional challenge of conflicting terminology is explored before examples of how spiritual needs and care are charted/documented are provided.

L. Ross
School of Care Sciences, Faculty of Life Sciences and Education, University of South Wales, Pontypridd, Wales

W. McSherry (✉)
Department of Nursing, School of Health and Social Care, Staffordshire University, Staffordshire, UK
e-mail: W.McSherry@staffs.ac.uk

© The Author(s) 2020
S. Peng-Keller, D. Neuhold (eds.), *Charting Spiritual Care*,
https://doi.org/10.1007/978-3-030-47070-8_6

97

1   Spirituality Within the UK National Health Service (NHS): A Core Concept

As early as the fifth century BC, Hippocrates recognised the importance of both the body and the soul in health. Historically, in Western medicine, the body and soul were inseparable with the sick being cared for as whole beings, body, mind and spirit. An example is the Knights of St. John, who set up a hospital to care holistically for sick pilgrims in eleventh-century Jerusalem. It is now known as the St John Ambulance, the UK's leading first-aid charity. Focus on care of the whole person continued to be at the heart of the newly formed NHS in 1948, where hospital chaplains were employed as specialist spiritual care providers. They are still employed in this capacity today. The NHS Constitution puts 'the patient [...] at the heart of everything the NHS does' (Department of Health & Social Care 2015, 3). Spiritual needs and care are important to people when faced with life's challenges such as illness (Selman et al. 2018). Healthcare delivery is to be evidence driven (www.nice.org.uk/guidance [n.d.]), and evidence shows that spiritual wellbeing is positively associated with quality of life and fosters coping mechanisms (Koenig et al. 2012; Steinhauser et al. 2017). Spirituality, therefore, features in healthcare policy at world level (e.g. WHO 2002), within Europe (especially in palliative care, e.g. EAPC (www.eapcnet.eu/eapc-groups/task-forces/spiritual-care [n.d.])) and within the UK (e.g. Welsh Government 2015; NICE no date).

2   Who Provides Spiritual Care?

Spiritual care is provided by 'healthcare staff, by carers, families and other patients', whilst 'chaplains are the specialist spiritual care providers' (UKBHC 2017, 2).

2.1   Healthcare Staff

All healthcare staff can provide spiritual care. Nurses, however, are in a unique position as they are the only healthcare professionals who are with the patient 24/7. Nurses act as gatekeepers to other services, including chaplaincy, and they advocate on the patient's behalf. Spiritual care is part of their holistic caring role (ICN 2012), and they are expected to be competent in assessing spiritual needs and in planning, implementing and evaluating spiritual care (NMC 2018). However, they face a number of challenges.

2.1.1 What and How to Assess and Document?

Nurses are expected to document the care they give and the decisions they make, but there is considerable variation across the UK in how this is achieved in practice. In Wales, the need for a unified approach to assessment in nursing has been identified, and a single streamlined assessment document has been developed and is currently being piloted in acute hospital settings across the country (e-nursing project). Assessment of spiritual care needs is part of this document.

Although spiritual care is part of the nurse's role which nurses are required to document, evidence suggests that this may not happen routinely. For example, a survey conducted by the Royal College of Nursing of the UK in 2010 found that only 3 out of 139 (2.2%) respondents said they used a formal spiritual assessment tool; a similar study in Australia reported 18 out of 191 respondents (26%) using a formal tool (Austin et al. 2017). Instead, respondents said they identified spiritual needs informally by listening and observing.

2.1.2 More Education?

A growing body of international evidence suggests that nurses feel unprepared for spiritual care and want further education to enable them to be more confident and competent in this aspect of their role (RCN 2010). Lack of education, first identified by Ross (1994), appears to be the greatest barrier to nurses assessing/screening for spiritual needs and recording spiritual care. McSherry reported a similar finding in 1997; 71.8% ($n = 394$) of nurses in his sample said they felt unprepared and wanted further education.

In the late 1980s and 1990s, nurse education in the UK underwent a major restructuring, shifting from colleges and schools of health into the higher education sector with a change of emphasis from the apprentice-style training to a greater focus on academic study. Around the same time, there was an explosion of interest in the spiritual dimension of healthcare. One would have thought that these two significant changes would have resulted in nurses feeling more prepared for spiritual care; however it seems that little has changed. A survey conducted by the RCN in 2010 reported 79.3% of respondents still feeling unprepared for this aspect of their practice, and 79.9% calling for further education. The RCN responded by producing guidance in the form of a pocket guide and an on-line educational resource (RCN 2011). The EPICC Project 2016–2019 (www.epicc-project.eu) is the most significant response to date. Over three years, 31 nurse/midwifery educators from 21 countries co-produced a set of spiritual care competencies which are shaping nurse/midwifery undergraduate curricula across Europe. A toolkit provides teaching and learning activities to enhance competency development. In some countries, for example, Wales, the competencies are being used to shape curricula of other healthcare professions (e.g. professions allied to medicine) and ancillary staff. It will be interesting to see if there is any change in nurses'/midwives' preparedness for spiritual assessment and care should the RCN survey be repeated when the first graduates from these new programmes become qualified nurses and midwives in five or six years' time.

2.2   Chaplains/Spiritual Care Givers

Healthcare chaplains are the specialist spiritual care providers in the NHS in the UK. Traditionally they were ministers of religion from the Christian faith who were paid for by the NHS but managed through their churches. Hospital chapels were built where church services and rituals such as baptism and communion were conducted. However, societal changes in the last two decades, whereby society has become increasingly secular and multi-cultural, have meant that chaplaincy has had to evolve to meet the changing needs of the people it serves. Scottish health policy introduced in 2002 (HDL 76) saw the remit of chaplains broaden to include care of 'spiritual' as well as 'religious' needs of people of all faiths (of which there are now many) and no faith, and this broader remit has filtered throughout the UK. Recent years have seen even greater role change as chaplains respond to the needs of the people and organisations they serve. Roles have diversified to include, for example, specialist roles, interdisciplinary team working, training and education, training and management of volunteers, research and evaluation including audits and increasing engagement with stakeholder groups including third sector organisations. This means that chaplains are working in increasingly varied ways in increasingly diverse settings. These changes have presented a number of challenges for chaplains/spiritual care givers.

2.2.1   What to Call the Service?

Many hospitals have changed their name from 'chaplaincy department' to 'department of spiritual care' to reflect the wider remit of the service they provide.

2.2.2   Who Chaplains Are?

The requirement to be a minister of religion may no longer be appropriate in a society which is becoming increasingly secular and where religious practice is in decline. Some hospitals now employ humanist or 'secular' chaplains to lead their departments. Many larger hospitals will have teams made up of different faith leaders to represent the multifaith profile of the communities they serve.

2.2.3   What Qualifications Are Needed to Be a Chaplain?

If it is no longer necessary to be a minister of religion, then what training is needed, and how does this differ from that, say, of a psychologist or a psychotherapist? There is currently no common training programme to become a chaplain, although in Scotland a new postgraduate qualification in spiritual care is in development. Chaplaincy volunteers, who may have no qualifications, are heavily relied on in many hospitals, raising questions about the appropriateness of this and the training they may need.

2.2.4   Who Should Be Paying for a Service?

As chaplaincy offers religious support within its remit in a society where religious practice is in decline, some question whether the NHS should continue to pay for it. The National Secular Society, for example, wants funding for hospital chaplaincy to stop and be reinvested in other ways, such as for the purchase of equipment.

2.2.5   Whether Chaplains Are Professionals and Who They Are Accountable to?

Chaplains work alongside other professional groups such as doctors, nurses and professions allied to medicine. All these professions have a recognised education programme, a code of conduct and a regulatory body which sets criteria and standards that must be reached and maintained to ensure safe practice. There are no such mandatory requirements for hospital chaplains.

Since 2007, all whole time chaplains in Scotland became employees of the NHS rather than the church. This became common practice also in the UK. Chaplains have, therefore, come under greater scrutiny regarding the effectiveness, cost-effectiveness, safety and quality of their practice. In response to this change in governance, a number of associations were set up to which chaplains could belong, some according to country, others by specialty (Table 1). However, the UK Board of Healthcare Chaplaincy (UKBHC) is the only body offering professional registration for chaplains. Accredited by the Professional Standards Agency, the UKBHC

Table 1 Examples of chaplaincy associations	College of Healthcare Chaplains
	Chaplaincy Accreditation Board (Ireland)
	Healthcare Chaplaincy Board (Ireland)
	Association of Hospice & Palliative Care Chaplains
	Northern Ireland Healthcare Chaplains Association

Table 2 Examples of chaplaincy guidelines, standards, competencies and capabilities in the UK

Document	Country	Year
Standards for Hospice and Palliative Care Chaplaincy (AHPCC 2006)	UK	2006
Standards for NHS Scotland Chaplaincy Services (NES 2007)	Scotland	2007
Spiritual and Religious Care Capabilities and Competences for Healthcare Chaplains Bands (or Levels) 5, 6, 7 & 8 (UKBHC 2017)	UK	2017
NHS Scotland. Spiritual care & chaplaincy	Scotland	2009
UKBHC Standards for Healthcare Chaplaincy Services (UKBHC 2009)	UK	2009
NHS England. NHS Chaplaincy Guidelines 2015	England	2015
Standrad for spiritual care services in Wales 2010 (WG 2010)	Wales	2010

provides a code of conduct, standards, competencies and capabilities for chaplains as well as a fitness to practice process. Although registration is currently voluntary, the intention is that this will become mandatory in the near future. Table 2 shows the chaplaincy standards, competencies and capabilities that currently exist in the UK, although those provided by the UKBHC (2009) are generally regarded as the most up to date. Additionally, chaplaincy guidelines have been produced in England (Swift 2015).

3 In Pursuit of Examples of Charting/Recording/ Documenting/Assessing in the UK

It was against this complex backdrop that we sought to identify examples of charting/recording/documenting/assessing of spiritual care in the UK, and our next challenge emerged, that of conflicting terminology.

3.1 Conflicting Terminology

A simple search of the CINHAL full-text database was undertaken in autumn 2018 using a range of search terms as follows:

- 'charting and spirit*' yielded a total of 19 records (reduced to 1 when 'and' was removed). Most pertained to charting of spiritual care within the fields of chaplaincy and parish nursing.
- 'documenting and spirit*' yielded 39 results (reduced to 9 when 'and' was removed), but the majority were unrelated to the documentation of spiritual care as defined for this book and were from a broad spectrum of disciplines.

– 'recording and spirit*' yielded 59 records (reduced to 8 when 'and' was removed), but again most were not directly relevant to the topic of this book.
– 'care plan and spirit*' yielded the highest return, 259 results. Concrete examples of tools were not given in these articles.

This simple search highlights issues around the use of language and how the terms 'charting/documenting/recording' may be interpreted differently across healthcare professions, countries and cultures when applied to meeting the spiritual, religious or pastoral needs of a person.

This brief analysis highlights that there is no internationally recognised universal term for charting/recording/documenting of spiritual care. In the UK, the word 'charting' would tend to relate to the recording of a patient's vital signs or fluid balance on a chart. 'Records' could apply to all documents that pertain to a patient whether in paper or electronic formats. The term 'care plan' is well recognised in nursing and achieved the greatest number of hits in the above search, picking up articles relating to the recording of spiritual issues by nurses.

This search highlights that different professions are involved in spiritual care and that terminology may be discipline specific. We now give examples of charting/documenting/assessing/recording from nursing and chaplaincy.

3.2 Examples of Charting/Documenting/Recording/Assessing by Nurses in the UK

We have already identified that professional groups, such as nurses, are required by their regulatory body to keep a record of what they do (NMC 2015) and that the actual format of that record will vary greatly. What is less clear, however, is what exactly nurses record about spiritual needs and care. We provide three examples.

3.2.1 Care Planning in Mental Health in England

Walsh et al. (2013) conducted a comparative study evaluating the efficacy with which care plans capture and make use of data on the spiritual and religious concerns of mental health service users in one UK Health and Social Care Trust. The findings reveal that (a) the importance that many service users accorded to spirituality and religion was not reflected in the electronic records, (b) that some information was wrong or wrongly nuanced when compared with the patient's self-description and (c) that service users themselves were often mistaken regarding the type and quality of information held on record. Walsh et al. (2013, 161) conclude that this may be because 'the majority of Care Coordinators are unable to see the relevance of spiritual or religious concerns, or feel incompetent to record them faithfully'.

3.2.2 Care Planning in Mental Health in Wales

In Wales, it is a legal requirement that people being treated for mental health problems have a Care and Treatment Plan (CTP). There are eight domains within the CTP covering people's medical needs, financial needs, accommodation needs, etc. Domain 7 asks about a person's 'social, cultural and spiritual' needs. In 2016 we conducted a pilot study in three of the seven Health Boards in Wales to find out what was written in Domain 7 of the CTP and to explore staff's experience of completing it (Fothergill 2018).

What was written in the CTP? Assessments were mainly carried out by care co-ordinators who were usually mental health nurses. We typed out verbatim what was written in Domain 7 from 150 CTPs (a mix of community and hospital records, 50 from each Health Board). Two researchers independently conducted a content analysis identifying 11 themes which were subsequently reduced to 8 (small themes were merged). The key findings are shown in Table 3.

Mainly social needs were recorded, e.g. involving people in meaningful activities (memory books, hobbies, etc.) and maintaining social connections with family and staff. The spiritual needs recorded were mainly of a religious nature, for example, facilitating attendance at religious services and listening to religious music.

Staff's experience of completing the CTP: We ran two focus groups in 2016, one with seven nurses and the other with four nurses who were care co-ordinators, experienced and newly qualified mental health nurses from both hospital and community. Sessions were tape recorded and transcribed and themes were identified. The staff stated that Domain 7 was the least frequently completed and that they found it the most difficult to complete. They found spiritual needs difficult to articulate and therefore difficult to assess, and they had not had any training which they felt would be beneficial. When we encouragingly talked about spiritual care, they saw it as something broader than religion; it was about identifying what gives a person meaning/peace/contentment, what puts a smile on someone's face or what gives hope, but they did not document this as spiritual care.

Table 3 Themes: Domain 7 (social, cultural, spiritual) ($n = 150$)

	Number ($N=$)	Percentage (%)
Meaningful activities & interests	85	56.7
Social connections with family, friends	73	48.7
Social connections with staff	59	39.3
Religion/religious beliefs	44	29.3
Supporting family and carers	19	12.7
Person-centred care	13	8.7
Considered but lacking in detail	13	8.7
Culture, e.g., Welsh language	11	7.3

3.2.3 A Spiritual Assessment Model for Nurses

Nurses are key to patients having their spiritual needs assessed, but evidence suggests they feel unprepared for this part of their role and often see it as an extra task on top of an already heavy workload. As a result, patients are at risk of their spiritual needs being ignored. We, therefore, developed a pragmatic spiritual assessment model, the '2 question spiritual assessment model' (2Q-SAM; see Fig. 1 and Ross and McSherry 2018), to help nurses to undertake a spiritual assessment that may not require any extra time. Instead, the model encourages nurses to give care that is inherently 'spiritual' because it addresses what is most important to the patient at any point in time (integrated rather than an 'add-on'). It may even save time (addressing the prudent healthcare agenda), as time will not be wasted on unnecessary tasks. The two questions are 'what's most important to you?' and 'how can I help?'. Additional benefits are that care is co-produced, person-centred and needs led. The model is currently being field tested.

4 Examples of Charting/Documenting/Recording/Assessing by Chaplains in the UK

We struggled to find any concrete examples of how chaplains record what they do from on-line searching. This led us to question whether there is any such requirement for chaplains to record what they do in the UK. We did this by:

(1) Conducting a key word search (search terms 'chart', 'document' and 'record') of four documents listed in Table 4, one each from the UK, Scotland, England and Wales
(2) Contacting chaplains personally known to us, in Scotland, England and Wales to ask for examples from their practice

4.1 Conducting a Key Word Search

Table 4 shows the results of the search.

It is clear from Table 4 that chaplains are required to keep some sort of record of what they do but exactly what information should be included, and the format for that is not stipulated.

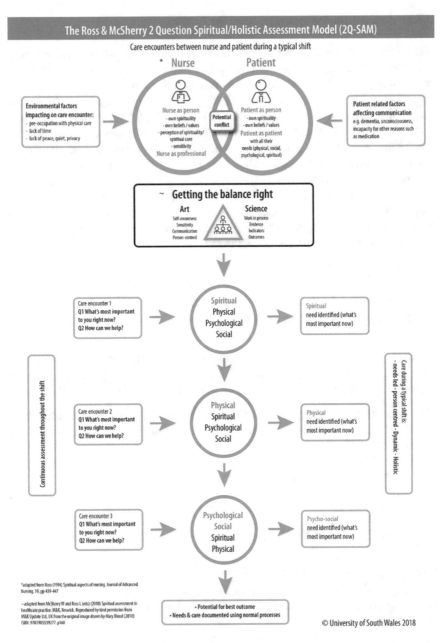

Fig. 1 The '2 question spiritual assessment model' (2Q-SAM) (Reprinted from Ross, L., McSherry, W. (2018). Two questions that ensure person-centred spiritual care. Nursing Standard [Internet]. https://rcni.com/nursing-standard/features/two-questions-ensure-person-centred-spiritual-care-137261, with permission from University of South Wales. Copyright © 2018 University of South Wales. All rights reserved)

Table 4 Key word search results from selected chaplaincy documents

Search term	Result
NHS Chaplaincy Guidelines 2015 (England)	
'chart'	0
'document'	0 other than referring to written documents such as reports and papers
'record'	'To ensure safety, accountability and continuity of care chaplains should maintain a record of work in a locally agreed format and in accordance with NHS policies for record keeping' (9)
	Several references to the need to record patients' religion or belief in hospital/clinic records (16, 24)
Standards for NHS Scotland Chaplaincy Services 2007 (Scotland)	
'chart'	0
'document'	There should be a system for accurate documentation and referral for those who wish to request a visit from a chaplain/spiritual care giver or chosen faith representative (9)
'record'	Spiritual care services should have
	6.a.3 access to patient information systems for providing and facilitating appropriate spiritual or religious care and recording information and interventions (9)
	For staff support under 'self-assessment'
	Are incidences (not content) of support recorded? (e.g. a diary/log noting the time spent and whether professional or personal. No name or content need be recorded, preserving confidentiality) (21)
Standards for Healthcare Chaplaincy Services 2009 (UK)	
'chart'	0
'document'	Identical to NHS Scotland (2007)
'record'	Identical to NHS Scotland (2007)
Standards for Spiritual Care Services in the NHS in Wales 2010 (Wales)	
'chart'	0
'document'	Identical to NHS Scotland (2007)
'record'	Identical to NHS Scotland (2007)

4.2 Contact with Chaplains

Chaplains from one country, who were together for another purpose, were asked about examples of how they document what they do, but they were reluctant to share this; that door was firmly closed.

One lead chaplain from another country pointed out that, whilst there is a requirement to keep a record, any attempt to develop a standard format would be problematic because of the extent of local variation in chaplains' practice. In his organisation chaplains record their visits directly in the patient's electronic record, alongside that of other healthcare providers (HCP) in line with a protocol established following an investigation.

A lead chaplain from another country said that chaplains have each developed their own personal ways of recording what they do, producing their own forms, systems or spreadsheet to track contacts and visits.

We approached five other chaplains personally known to us and had responses from four of these. They were willing to share how they record what they do, and three were willing for their documents to appear in this book. All wished to remain anonymous. The overwhelming concern seemed to relate to Information Governance, particularly in relation to breaching patient confidentiality. Additionally, many chaplains simply were not convinced of the need to record anything.

4.2.1 An Example from a Chaplain in Scotland

The Health Board has one lead chaplain, a chaplaincy co-ordinator and a number of community volunteers. No assessment tools are used. Rather, patients identified on the hospital computerised system as belonging to a particular religion, or wishing to see a chaplain, are visited. The conversation between patients and the chaplain dictates what care is given. The chaplain keeps a record of visits in a notebook. This is not generally shared with any other members of the multidisciplinary team (MDT), except in the palliative care unit at whose MDT meetings the Chaplain has traditionally been welcomed.

4.2.2 An Example from a Chaplain in Wales

The Health Board has several hospitals, and each one has its own approach to documenting. In one hospital, when a referral is received, a chaplain will visit a patient, and the conversation will be guided by the outline in screenshot 1. Relevant information will be recorded in the Chaplaincy Patient Care Plan (screenshot 2) using 'The Graph' (screenshot 3) to prioritise patients' needs (Figs. 2, 3 and 4).

4.2.3 An Example from a Chaplain in Wales

This Health Board has four hospitals, and the chaplains in one hospital developed a tool for recording their contact with patients. It records how the referral was made, what was done (whether spiritual or pastoral care, Holy Communion or prayer was given) and whether another visit is planned and leaves space for further comment (screenshot 4) (Fig. 5).

4.2.4 An Example from a Chaplain in England

It is clear when talking to several lead chaplains in England that they are using a range of different methods to record and capture key information regarding referrals and contact with patients and their carers. Table 5 shows the types of information being recorded in one NHS Trust in various ways including notebooks, Excel spreadsheets and other electronic recording systems.

How does a conversation work?

Every conversation is different, but there are some common themes, and some mistakes to avoid.

This framework aims to help us understand how a conversation flows. It is not that you have to ask all these questions (that would almost certainly be insensitive). This framework may help you to:

- Think why you are asking certain questions
- Listen better to people's stories and the emotions within them
- Decide when and how to offer prayer, and what to pray for/about.

Stage	The conversation	The rationale
Beginnings	• Hello my name is……, I'm a chaplaincy volunteer. What's your name? • How are you getting on? • Have you been here for long? • • Where do you live usually? • Do you have any visitors?	Introducing yourself. Ascertaining their correct/preferred name. Opening question – not too specific. Length of time in hospital may lead to further questions: have you settled? Are you bored? Are you in shock? Ascertains whether they are close to home. Ascertains whether they have social support.
Ending (short contact)	The conversation may end here if: • The person appears tired or unwilling to talk • The person tells you they are not interested. Do not attempt to prolong a conversation. Smile. Ask if they need anything. Leave.	People demonstrate their unwillingness to talk by: • Not giving any attention or eye contact • Giving clipped responses Prolonging conversations achieves nothing except frustration!

Middle	• What brought you into hospital?
	How did this illness start?
	What have the doctors told you?
	How long are you expecting to stay?
	How are you feeling about it?
	• Do you have any family?
	• Who else is affected by your illness?
	Has your work/responsibilities suffered?
	Have you got any dependents?
	How do you feel about them?
	• How strong are your relationships?
	Have you been bereaved?
	Do your family/friends care about you (enough)?
		We do not ask questions just for information, and certainly not to form judgements or express opinions. The aim of all these questions is to allow people to:
		  a) tell their story
		  b) express their emotions.
		  c) Get help if needed.
		
		We then respond to emotions with empathy which usually soothes.
		
		We will want to remember the names of family and friends in case the patient allows us to pray for them later.

	Are any relationships causing you pain? • Are you satisfied with the care in hospital? *Are you being treated with respect/kindness?* *Do you understand what is happening?* *What will happen when/if you go home?* *How are you feeling about the future?* • What is helping you to keep going? • Where do you get your hope from? *Do you have any links with a church or faith group?* *Have you had links in the past?* *Do you want any religious support from us: to come to chapel, to be prayed for, to be visited by a chaplain?* *Do you want us to contact anybody in your faith group?* *How are you feeling about your relationship with God/church?*	If the patient has concerns we will want to check that these are being followed up by the most appropriate person. We will not assume that everything the patient says is absolutely correct, but we will assume it is the truth as they see it. This is some kind of spiritual assessment. Without asking people directly about their religion or belief, we are tentatively enquiring into their sources of support. This treads the fine balance between assuming too much and offering too little. If they ask for some support, a) write it down, b) pass it on, c) check it out with the chaplains, if you are unsure.
Endings	• Would you like me to pray for/with you? *Would you like me to pray here with you now or for you in the chapel later?* • Do you think it would help to talk further? I'm not going to be here again for a few days. Would you like me to ask a chaplain to visit you? • Is there anything else that you need? • Thank you for talking to me. It has been good to meet you, • *I hope you are home/better soon.*	This is what we do! It is always fair to ask, unless the patient has already indicated that they are atheist (in which case it is rude and insensitive to offer). When we pray we ask God to help resolve the painful emotions and lay before Him the areas of concern. It is important to end a conversation well, and to let patients know where they can find further support. You will probably need to inform patients that chaplains are available to listen and support (just as you have done); they are not going to preach or embarrass them. Obtain consent for any referrals. Let the patient know what you are proposing to do with any information they may have entrusted to you.

Afterwards make sure that you record any new referrals in the referral book.

Fig. 2 Conversation of chaplain in Wales, guided by the outline shown above

Chaplaincy Department
Patient Care Plan

Name of Patient		Ward:	Communicant Yes / No
Referred By:		Denomination(if any)	

Date of Visit	Comments	Signature

Fig. 3 Chaplaincy patient care plan

The numbers 1-5 represent the order of priority of need, Number 1 high and number 5 low.

Be aware of the needs and after studying the questions, put a tick in one or more of the squares. A clearer picture will emerge of the needs to be addressed and the order of priority.

The Graph

Needs	1	2	3	4	5
1. Being valued					
2. Meaning					
3. Hope					
4. Emotions					
5. Dignity					
6. Truth and Honesty					
7. Language					
8. Death/Dying					
9. Religion					
10. Culture					

Fig. 4 Prioritisation of patient's needs

Chaplaincy Spiritual Care Service 2015
SIGNIFICANT PATIENT CONTACT/COMMUNICATION SHEET

Week beginning_____ Week_____

NAME	WARD	ROUTE					OUTCOME					COMMENTS
		Staff	Self	Relative	Minister	Chaplain	Spiritual	H.C.	Pastoral	Prayer	Visit again	

Fig. 5 Tool for recording contact with patients, Wales

Table 5 An example of some of the information that chaplaincy departments are recording after a patient referral and contact

Name	Hospital No	Religion	Call	Referral	1st seen	Total visits	Latest visit	1st Holy Communion	Baptism	Blessing (anointing)	Date Discharge
Notes											

5 Conclusion

We have been unable to identify any nationally recognised approaches by chaplains to assess and document spiritual needs and care in the UK. Informal enquiries with some chaplains in England, Scotland and Wales indicate that formal approaches to assessment, such as the use of tools, are not generally used. Instead, spiritual needs are assessed informally through conversation with the patient. If and how information obtained from these conversations is documented or shared (and with whom) is also difficult to determine. It seems that some chaplains do not document or share at all, whilst others have developed their own ways of keeping records of visits within their own teams. Outside of palliative care, chaplains may not be included in MDT meetings. In many places they are unable to access or contribute to patients' medical notes, despite supposedly being 'expected to take their place as members of the multiprofessional healthcare team' (UKBHC 2017, 2). The audit of chaplaincy services appears to be equally piecemeal.

The assessment of spiritual needs and documentation of spiritual care by nurses is equally vague. Whilst admission forms generally ask about a person's 'religion', this is frequently left blank. If and how spiritual care needs are included within care plans is also unclear. The examples we have provided give some insight into how this is being done in some mental health settings in Wales. A PhD study starting shortly will engage key stakeholders in developing unified spiritual assessment and documentation processes that are fit for purpose across Wales.

We conclude that there is no standardised means of assessing and documenting spiritual needs and care in the UK and that this is unlikely to become a reality until the many complex challenges we have outlined in this chapter are addressed. A new study on the professionalisation of chaplains across the UK, if fully funded, may make a start on addressing some of these complexities.

Commentary

David Neuhold (✉)
University of Zurich, Zurich, Switzerland
e-mail: david.neuhold@unifr.ch

Spiritual care does not exist in vacuum. The framework conditions of historical circumstances and developments in the history of religion and spirituality as well as political-cultural framework factors are decisive in determining how religious and spiritual facets are interpreted and what significance is attached to these issues in their respective institutional framework – issues that exceed the bio-psycho-social triangle in healthcare.

1 The Importance of Cultural Background and Political Setting

The contribution by Linda Ross and Wilfred McSherry shows that in the United Kingdom the integration of the spiritual dimension in a hospital setting is tough. And this despite the fact that the spiritual dimension was recognized in the British health system (NHS) as early as 1947. As an example, McSherry and Ross make this difficult situation clear by the fact that humanist groups compare spiritual care and infrastructural conditions in financial terms and aim them at one another. Humanists present spiritual care providers and hospital infrastructure as competing for financial resources. As if both were a straight contrast, a zero-sum game. Thus, on the one hand, there are strong tendencies towards the privatization of the religious-spiritual factor and the religious-spiritual care provision; on the other hand, even in the United Kingdom there are regionally different approaches. In Wales, for example, there is clearly a greater sensibility for the spiritual dimension of the patient in the context of 'care planning in mental health'. This chapter also draws attention to such differences.

A crucial area in the field of spiritual care is that of training. This is exactly where appropriate competences are acquired and sensitivities are instilled. McSherry and Ross bring this out skillfully in identifying gaps in the training of nursing staff. They find out that many nurses have not been prepared for 'spiritual care' or feel incompetent to cope with specific demands. Both of them, McSherry and Ross, are therefore careful to include spirituality more centrally in the training processes. They themselves have launched EPICC ('Enhancing Nurses' and Midwives' Competence in Providing Spiritual Care Through Innovative Education and Compassionate Care'), a European-wide project in this context. This research project aims to analyze the shortcomings identified and take steps to remedy them.

2 The Decisive Intersection of Spiritual Care and Nursing

The first point of contact for spiritual concerns and needs in the hospital context is often not the pastor, chaplain or spiritual care provider, but the nursing staff. Spiritual care is also their responsibility. Ross and McSherry focus their attention on this, as both also have a background in nursing. This aspect makes their contribution to the present volume unique and lends it a special authority. Ross and McSherry have also developed important tools for this first contact in the hospital, such as the 2Q-SAM tool. From the spiritual point of view, it is now crucial to work closely with the nursing staff and to find a suitable and understandable common language. It is highly incumbent upon those working in the field of spiritual care to establish and develop a clear and translatable terminology. The factor of education and training comes into play again. In general, Ross and McSherry are particularly sensitive to linguistic terms and terminology. This sensibility for language is particularly relevant in the medical context if one is committed to providing the best care for patients.

3 An Aside Concerning Islamic Spiritual Care in the United Kingdom

One aspect that was not directly addressed in the chapter is the growing area of Islamic pastoral care in hospitals. In the United Kingdom, as in the Netherlands, this is becoming ever more important. The number of Muslim citizens in these countries is steadily increasing. The importance of the issue of Muslim spiritual care was recently pointed out by Dilek Uçak in her contribution to a book on spiritual care in a global context (Simon Peng-Keller/David Neuhold, Spiritual Care im globalisierten Gesundheitswesen, Stuttgart 2019, p. 207–230). Uçak compares spiritual care in European countries like the United Kingdom with that in Turkey and Iran. While in Turkey the religious-political dimension is becoming more important, in Iran spiritual care by the nursing staff is further developed. The Turkish and Iranian contexts could provide Ross and McSherry with new starting points for comparison. In connection with the documentation of clinical pastoral care, it should be mentioned that spiritual care work is also documented in Turkey. The spiritual care giver is required to fill in three text fields, providing a general description of each case, along with measures and results.

References

Association of Hospice and Palliative Care Chaplains. 2006. Standards for Hospice and Palliative Care Chaplaincy.

Austin, P., R. MacLeod, P. Siddall, W. McSherry, and R. Egan. 2017. Spiritual care training is needed for clinical and non-clinical staff to manage patients' spiritual needs. *Journal for the Study of Spirituality* 7 (1): 50–53.

Department of Health and Social Care. 2015. *The NHS constitution.* London: Crown.

European Association for Palliative Care. n.d. EAPC task force on spiritual care in palliative care. https://www.eapcnet.eu/eapc-groups/task-forces/spiritual-care. Accessed 20 Dec 2018.

Fothergill, A. 2018. *Exploring how the spiritual needs of dementia patients are addressed within Care and Treatment Plans (CTPs) in three Health Boards (HBs) in Wales*, Report. Pontypridd: University of South Wales.

International Council of Nurses (= ICN). 2012. *Code of ethics for nurses.* Geneva: ICN.

Koenig, H., D. King, and V. Carson. 2012. *Handbook of religion and health.* New York: OUP.

McSherry, W. 1997. *A descriptive survey of nurses' perceptions of spirituality and spiritual care.* Unpublished MPhil thesis. Hull: The University of Hull.

McSherry, W., and L. Ross. 2002. Dilemmas of spiritual assessment: Considerations for nursing practice. *Journal of Advanced Nursing* 38 (5): 479–488.

———. 2010. *Spiritual assessment in healthcare practice.* Keswick: M&K Publishing.

National Institute for Health and Care Excellence. 2011. *End of life care for adults: Quality Standard* 13. Available at www.nice.org.uk/guidance/qs13/documents/qs13-end-of-life-care-for-adults-quality-standard-large-print-version2. Accessed 27 Mar 2019.

National Institute for Health and Care Excellence (= NICE). n.d. www.nice.org.uk/guidance. Accessed 19 Mar 2019.

NHS Education Scotland. 2007. *Standards for NHS Scotland Chaplaincy Services 2007.* Edinburgh: NES.

NHS HDL 76. 2002. Available at www.sehd.scot.nhs.uk/mels/hdl2002_76.pdf. Accessed 25 March 2019.

Nursing and Midwifery Council. 2018. *Standards of proficiency for registered nurses.* London: NMC.

Nursing and Midwifery Council (= NMC). 2015. *The code.* London: NMC.

Ross, L.A. 1994. Spiritual aspects of nursing. *Journal of Advanced Nursing* 19: 439–447.

Ross L., and W. McSherry. 2018. *Two questions that ensure person-centred spiritual care*. Available at: rcni.com/nursing-standard/features/two-questions-ensure-person-centred-spiritual-care-137261. Accessed 27 March 2019.

Royal College of Nursing. 2011. *Spirituality in nursing: A pocket guide*. London: RCN.

Royal College of Nursing (= RCN). 2010. *Spirituality survey 2010*. Available at www.rcn.org.uk/professional-development/publications/pub-003861. Accessed 5 Dec 2018.

Selman, L., L. Brighton, S. Sinclair, I. Karvinen, R. Egan, P. Speck, et al. 2018. Patients' and caregivers' needs, experiences, preferences and research priorities in spiritual care: A focus group study across nine countries. *Palliative Medicine* 32 (1): 216–230.

Steinhauser, K.E., G. Fitchett, G.F. Handzo, K.S. Johnson, H.G. Koenig, K.I. Pargament, et al. 2017. State of the science of spirituality and palliative care research part I: Definitions, measurement, and outcomes. *Journal of Pain and Symptom Management* 54 (3): 428–440.

Swift C. 2015. *NHS chaplaincy guidelines 2015: Promoting excellence in pastoral, spiritual and religious care*. Available at hcfbg.org.uk/wp-content/uploads/2013/09/nhs-chaplaincy-guidelines-2015.pdf. Accessed 27 Mar 2019.

UK Board of Healthcare Chaplaincy. 2017. *Spiritual and religious care capabilities and competences for healthcare chaplains bands (or levels) 5, 6, 7 & 8*. Cambridge: UKBHC.

UK Board of Healthcare Chaplaincy (= UKBHC). 2009. *Standards for healthcare chaplaincy services*. Cambridge: UKBHC.

Walsh, J., W. McSherry, and P. Kevern. 2013. The representation of service users' religious and spiritual concerns in care plans. *Journal of Public Mental Health* 12 (3): 153–164.

Welsh Government. 2010. *Standards for spiritual care services in NHS Wales 2010*. Cardiff: Welsh Government.

———. 2015. *Health and care standards*. London: Crown.

World Health Organisation (= WHO). 2002. *WHOQOL-SRPB field test instrument*. Geneva: WHO.

Spiritual Care and Electronic Medical Recording in Dutch Hospitals

Wim Smeets and Anneke de Vries

Since the beginning of this century, a growing number of hospitals in the Netherlands have introduced electronic medical recording (EMR) systems. The focus in these systems is on recording medical and nursing data, but quite a few allow for data to be entered from other disciplines, such as spiritual care, as well. Crucially, data recorded in EMR are accessible to others – even if access can be controlled and restricted in various ways.

The use of EMR raises several important questions, which spiritual caregivers have been trying to answer from the start. The present article provides a survey of this discussion as it has been conducted in our country and in which the following issues have predominated: to record or not to record, legal issues, and the "why, what, and how" of recording. Using a case in point, we will illustrate our own recording practices at Radboud University Medical Center (henceforth: Radboudumc). Finally, we will outline some perspectives for the further development of EMR.

1 To Record or Not to Record?

On the question of whether or not a spiritual caregiver should use EMR to record patient-related data, we find that some colleagues are vehemently opposed to the whole idea, while others are working on it with enthusiasm. The opponents are typically cultivating a position of non-involvement in the therapeutic process. Elsewhere we have characterized this position as isolationist, outlining its benefits as well as its risks (Smeets, Gribnau and Van der Ven 2011). The main benefit is that it safeguards the "sanctuary" to which each patient is constitutionally entitled; the main risk is

W. Smeets (✉) · A. de Vries
Radboud University Medical Center, Nijmegen, The Netherlands
e-mail: Wim.Smeets@radboudumc.nl

© The Author(s) 2020
S. Peng-Keller, D. Neuhold (eds.), *Charting Spiritual Care*,
https://doi.org/10.1007/978-3-030-47070-8_7

spiritual care becoming irrelevant. A related issue here is confidentiality, which we will discuss below. The opponents' principal argument against EMR is that it is incompatible with the confidential nature of spiritual care. They also see it as disproportionally time-consuming.

By contrast, colleagues working with EMR typically take a position of involvement in the therapeutic process, alongside other healthcare professionals. We have characterized this position as assimilatory, with participation in the multidisciplinary therapeutic process as its main benefit and loss of identity as its main risk. From this position, the principal argument in favor of recording is that it may be in the patient's interest to share information with other professionals involved in the therapeutic process (Smeets 2006, 120–144).

2   Legal Restrictions

From a legal perspective, the use of EMR generally, and by spiritual caregivers in particular, is not at all unproblematic. Safeguarding the patient's privacy is a central issue in the discussions about EMR and many other such systems in the Netherlands. A project by the Dutch government to develop a nationwide EMR was canceled due to massive resistance from the general public. Tellingly, quite a few doctors, on becoming patients themselves, are reluctant to consent to having their data recorded in EMR. In the Netherlands, the *Wet bescherming persoonsgevens* (Personal Data Protection Act 2000) has strengthened the patient's legal position, as it stipulates that data provided by the patient remain the patient's property and that the patient has the right to inspect recorded data at any time. Moreover, it is compulsory that the patient unambiguously consent to recording data in an EMR. Healthcare institutions are supposed to provide adequate procedures for this. As of May 2018, the Dutch Personal Data Protection Act has been replaced by the even stricter European General Data Protection Regulation EU 2016/679. This new regulation has a significant impact on the sharing of health-related information. Crucially, the client's explicit consent to documenting personal information about their health or to any other recording of personal data is now compulsory.

The sharing of recorded data is limited in principle to the healthcare professionals involved in a particular patient's therapeutic process. At the moment the *Wet beroepen individuele gezondheidszorg* (Individual Healthcare Professions Act) grants routine access to recorded data to the practitioners of some professions, but, importantly, not to spiritual care workers. In addition, Article 6 of the Dutch Constitution stipulates that each person has the right to practice his or her worldview, without any interference by others. That is why, traditionally, spiritual ministers have been bound not only to professional confidentiality but to ministerial confidentiality as well. The ministerial confidentiality regulations of the principal churches in the Netherlands are even stricter than most professional confidentiality regulations. Thus, as far as the use of EMR by spiritual caregivers is concerned, their legal status and their ministry would seem to constitute a double impediment

(Lammers/Smeets 2016). In practice, many EMR systems, including the one that is used at Radboudumc, do allow limited access to spiritual caregivers, legally underpinned by the hospital's legal experts. (It should be noted that patients are typically unaware of all this, assuming that spiritual caregivers are just normal healthcare professionals with normal access to the records.)

3   Why Use EMR?

With all the legal and ministerial strings attached to using EMR, why bother to even try? Actually, we think there are some valid reasons, pertaining to the patient's well-being, to healthcare generally, and to spiritual care as a profession.

As far as the patient is concerned, one might argue, first, that healthcare is an intrinsically multidisciplinary endeavor. Communication between care providers is essential so as not to confront the patient with fragmentation and contradictions in diagnosis and treatment. Some of this communication has to be in written form, in order to guarantee a reliable transfer of information. Second, the process of recording enables care providers to focus on what the patient needs right now. Third, EMR allows for the planning of the therapeutic process, as it not only records diagnosis and treatment so far but has medical appointment scheduling functionality as well. Fourth, the use of EMR allows the spiritual caregiver to influence the therapeutic process in such a way as to better serve the interests of the patient – even advocating them should they be contrary to those of the therapeutic team – for instance, by drawing attention to aspects of the patient's well-being that have been neglected so far.

As to the therapeutic process generally, one might argue, first, that EMR enhances mutual understanding between care professionals. Based on each other's findings and planning as recorded in the EMR, and through personal or multidisciplinary consultation, they develop a better understanding of each other's disciplines. Second, EMR allows for more efficiency in the therapeutic process, as it reduces overlap and duplications of work. And third, EMR facilitates a better assessment of the effectiveness of the treatments used to cure particular diseases.

Finally, there is an argument in favor of the use of EMR that specifically pertains to spiritual care, viz., that it has a positive effect on the professionalization of spiritual care. It makes us reflect on what we are actually doing and document it in such a way as to be transparent to others, i.e., to professionals from other disciplines as well as to fellow spiritual caregivers. The use of EMR induces the spiritual caregiver to take other professionals' perspectives. Which elements from our conversations with a patient are relevant to and need to be shared with other professionals to ensure good care for that patient, including their spiritual well-being – not only within the present context but in possible follow-up phases as well? Already, our EMR entries have started to play a role in multidisciplinary consultations. Even if long-term effects cannot be assessed as yet, it is our firm impression that other disciplines are beginning to develop a better understanding of the specifics of spiritual care as a profession. In addition, we are confident that the sharing between spiritual caregivers of information contained in EMR entries will help us realize a common standard of practice.

4 What to Record

There is a broad agreement among spiritual caregivers working with EMR that the data to be recorded in the system should concern the substance of the contacts, typically conversations, between spiritual caregivers and their patients. The basic strategy is to write a more or less condensed account of the conversation. This idiographic approach focuses on the individuality of the patient as a unique person with his or her own unique narrative. The main argument in favor of this approach is that it is the best way of doing justice to the patient as a person. Theoretical backing is provided by narrative approaches in theology, psychology, and, recently, medicine. From a person-centered care perspective, it has also been argued that reading these recorded conversations makes it easier for care providers to relate to individual patients, encouraging them to provide the best care they can.

A second approach has a rather more nomothetic character, as it focuses on similarities and differences between patients, for instance, through a classification of the topics addressed in a particular conversation. In the Netherlands, several classification proposals have been put forward. Some are theory-based, declaring and arranging topics in theoretically defined structures, while others are empirical, working with the topics that are actually raised by patients. An example of a theory-based proposal is the hermeneutic-diagnostic scheme developed by a group of spiritual caregivers in mental healthcare, using work by authors such as Erikson, Nagy, Capps, and Pruijser (Bos et al. 2003).

From a nomothetic-empirical angle, in the 1990s, a group of Roman Catholic healthcare chaplains in the Dutch province of Overijssel came up with a catalogue of 28 topics they found to be regularly addressed in pastoral conversations. The catalogue was used in a survey among all Dutch spiritual caregivers by the *Nijmegen Institute of Studies in Empirical Theology*, in cooperation with the *Netherlands Association of Spiritual Caregivers in Care Institutions, VGVZ*. The topics most frequently addressed were sickness, bereavement, life-changing events, relationships, suffering, meaning and appreciation of life, solitude/loneliness, worldview and faith, death, expectations, and the treatment received (Smeets 2009). Recently, a group of spiritual caregivers working in hospitals that use the HiX system have started work on a new catalogue of topics.

One benefit of this topic-based approach is in its user-friendliness. To record a conversation with a patient, the spiritual caregiver only needs to check the topics that have been addressed. In addition, it allows for various types of quantitative research into patient contacts, such as frequency rankings of topics. On the negative side, the selection of topics to be included in the catalogue is inherently problematic. In drawing up such a catalogue, one would typically start with concrete, real-life topics. There should not be too many of these, however, lest the catalogue become unmanageable. So the possibility remains of a topic being discussed which is not in the list. A solution to this problem could be to use more abstract "topical groups," as proposed by Van der Ven (1994, 10). Based on audio recordings of pastoral conversations, this author came up with ten "topical groups," viz., health, the self, social relations, societal problems, existential questions, religious questions, spiritual questions, moral issues, church-related issues, and pastoral issues. In our Nijmegen CPE training courses, we still use Van der Ven's topical groups in the analysis of pastoral conversations. However, the assignment of a topic to a topical group can still be controversial, and no categorization is everlasting: Who could tell the difference between "church-related issues" and "pastoral issues" nowadays?

The use of rather more formal categories appears to be a viable alternative. At Radboudumc we have drawn inspiration from George Fitchett's work in *Assessing Spiritual Needs* (2002). His model comprises seven thematic fields to categorize the topics that have been addressed:

- Faith and meaning
- Vocation, values, and responsibilities
- Impactful experiences and emotions
- Courage, hope, and growth
- Rituals and practice
- Community
- Authority and guidance

Dutch society is much more secularized than American society. Topics such as faith, authority, and guidance are not typically addressed in pastoral conversations. That's why we have tried to bring the categories more in line with the topics that do come up; in other words, we have created a more secularized version of Fitchett's model:

- Meaning and sources of inspiration
- Goals in life, values, and responsibilities
- Impactful experiences and emotions
- Courage, hope, and growth
- Rituals and practice
- Community and relationships

"Faith" has been replaced by "sources of inspiration," which may comprise other sources such as music, nature, or science. "Vocation" became "goals in life" – as it is very rare for people to define their work, life, or activities as the response to a vocation. "Community" is no longer exclusively understood as a religious community; but the term can also refer to other types of community, such as a village community, a commune, a group of good friends, or a family. Therefore, the topic "relationships" was added. The notions of "authority," as represented by the church, and "guidance" by a pastor do not occur as topics in our conversations with patients; they have therefore been eliminated. In our experience, the resulting six main topics fit our conversations much better than Fitchett's seven do (see Smeets/de Vries 2016).

Conveniently, an unlimited number of topics addressed in pastoral conversations can be assigned to this limited number of fields. Also, the model allows spiritual caregivers to identify their favorite areas of religion and/or spirituality, as well as any possible blind spots in their contacts with patients. Third, it communicates to other care providers that spiritual care is not just about religion and ecclesiastical matters but encompasses much broader areas. In our teaching experience, we find that other care providers are typically quite sensitive to this. Fourth, the fields just mentioned coincide with other well-known categorizations in religion and world-view, such as those by Cobb, Glock and Stark, and Smart (Smeets/De Vries 2016). The model thus provides an excellent base for comparative studies on an international scale. Finally, it provides a framework for recording the essential aspects of a patient's narrative as defined above, without recording the narrative as such. It can therefore be seen as a compromise between the idiographic and nomothetic approaches.

5   Recording at Radboud University Medical Center

Recording the substance of a conversation with a patient, essential as it may be, is not sufficient for the use of EMR by the spiritual caregiver – which is only modest, when compared to its use by nurses and doctors – to be relevant to the therapeutic process as a whole. What else is needed? Vandenhoeck (2007, 2009) has developed an integral recording method, building on the model by the American author A. Lucas, in *The Discipline for Pastoral Care Giving*. Vandenhoeck's method comprises the following elements:

– Why the contact took place
– Findings
– Interventions
– Results
– Follow-up

In our recording practice, we use this method in combination with our own version of Fitchett's model of spiritual assessment. The "findings" or, as we prefer to call this element, the "brief spiritual impressions," is where the substance of a patient contact is recorded. In our experience, the Fitchett assessment just mentioned is helpful in structuring the findings (see below, first case in point). Vandenhoeck stresses the need for factual recording, containing as few interpretations or other personal notes by the spiritual caregiver as possible. Thus, other care providers can get quick updates on the support by the spiritual caregiver, both on the findings and on the planned follow-up, and take these data into account in their treatment.

6   Cases in Point

At Radboudumc, the Department of Spiritual Care uses the EMR system *Epic* in the way just outlined, as the following cases in point may illustrate. A spiritual care-giver reports:

> One of my colleagues was called in for a baptism ritual on the neonatal ward. A 'spiritual care' order was placed in Epic. On the spiritual care page, my colleague reported: "Conversation this morning with the parents on baptism of X. Baptism took place at 13 hours, with parents and other family members present. The parents were very happy with it." Afterwards this colleague informed me, the order was transferred to me and I went to visit mother and child. I recorded our conversation as follows in Epic (in consultation with the mother).

A. *Why the visit took place*

Follow-up contact after baptism ritual, spoke with the mother.

B. *Brief spiritual impressions*

- *Meaning and sources of inspiration*: Being together with her child is what inspires Mrs. N. at this moment and what gives meaning to her life.
- *Goals in life, values, and responsibilities*: Talked about the balance between care for others and care for herself – the latter being as essential as the former.
- *Impactful experiences and emotions*: Mrs. N. would very much like to be in agreement with her husband with respect to the continuation of the treatment. She feels a tension.
- *Courage, hope, and growth*: The mother is cherishing hopes about the medical treatment of her child.
- *Rituals and practice*: Both parents have experienced the baptism ritual as supportive in a difficult moment.
- *Community*: Mrs. N. needs to be at home from time to time, in order to be in touch with her loved ones.

C. *Interventions*

Mainly been attentively listening to what Mrs. N. had to say.

D. *Results*

Mrs. N. had a sense of relief after the conversation.

E. *Follow-up*

Talked about possible further counseling, involving husband as well.

F. *(Multidisciplinary) referral – primary/secondary care*

Not applicable.

The referral category is not in Vandenhoeck's recording method. We added it with a view to our potential involvement in transmural care, specifically in providing spiritual care in the patient's home environment.

Notwithstanding all the modeling and structuring behind EMR, there is still considerable freedom when it comes to the actual recording of our conversations. Here is an example of a rather more succinct entry:

A. *Why the visit took place*

Patient had asked for a visit.

B. *Brief spiritual impressions*

Patient is emotionally tired and sad.

C. *Interventions*

Have listened to her experiences and sorrow and talked with her about end-of-life issues.

D. *Results*

Patient feels being heard, sense of relief.

E. *Follow-up*

Promised to visit her again Wednesday next week.

The perfect way to record data on patient contacts still needs to be worked out. On the one hand, the wording should be concise, as in the above examples, both to enable other care providers to quickly examine the records and to protect the patient's privacy. On the other hand, brevity carries a risk of misinterpretation, and the patient's individual narrative should not be obscured. Consultations within our team on this subject take place regularly and are likely to continue for some time.

7 Perspectives

In this contribution we have discussed our own use of EMR as implemented at Radboudumc, i.e., a system containing one spiritual care "page," which is accessible to all care providers involved in the therapeutic process. However, alternative setups are conceivable – each of them giving rise to their own legal, moral, and practical questions – such as having two pages, one accessible to all and the other with access limited to fellow spiritual caregivers, allowing more sensitive data to be recorded. Another option might be the inclusion of data from the realm of spiritual care in other care providers' pages. Also, in addition to EMR for hospital care, one might wish to have a system for outpatient care. At Radboudumc, some of these variants are already being worked on and we plan to report on them in due course. As mentioned already, the best way of recording data (in terms of wording) still needs to be worked out, possibly in consultation with patients and/or relatives as "producers" of the data and with the other care providers as "consumers."

8 Finally

Digital recording of spiritual care remains a contested issue in the Netherlands. At Radboudumc we feel that, on balance, the use of EMR in spiritual care is to be welcomed, as it serves several interests. It is in patients' interest that their concerns are remembered and taken into account beyond the conversation with the spiritual caregiver. It is in the other caregivers' interest to know the contribution of spiritual care and to integrate it into the therapeutic processes we are all working on together. Through EMR, healthcare managers can obtain a clear picture of what meaning and spirituality, as components of integral care, really entail. Finally, EMR helps spiritual caregivers to account for what they are doing and to study and improve their own practices. To ensure a stable future for our profession in healthcare, EMR seems, in our view, indispensable.

Commentary

David Neuhold (✉)
University of Zurich, Zurich, Switzerland
e-mail: david.neuhold@unifr.ch

1   Basic Reflections on Documentation

In their contribution, Smeets and De Vries provide a range of arguments to the effect that spiritual care workers should be involved in the documentation process and thus actively participate in EMR ("electronic medical records"). Nevertheless, they do not denigrate skeptical or even opposing positions. Rather, they think that there is a lot to be gained from listening to such critical voices. However, these "isolationist" voices, as Smeets and de Vries call them, are becoming quieter and quieter. The discussion of the article at our workshop at the beginning of 2019 confirmed this. There is also a generational gap. "Isolationist" voices often tend to be those of elder spiritual care workers.

According to Smeets and de Vries, it is important first and foremost that documentation does not become an end in itself: it should only be a means to an end. Secondly, documentation should not be excessively time-consuming (as it has already become in the Netherlands, and probably not only there). And thirdly, the written word cannot become a substitute for everyday interprofessional (oral) communication in the clinical context. As already clearly demonstrated in the book by Wim Smeets (*Spiritual Care in a Hospital Setting: An Empirical-Theological Exploration* from 2006), the author has an impressive knowledge of a number of concrete contexts. Great importance is attached to empirical investigation and to everyday conditions.

2   Importance of the "Narrative Approach"

An "idiographic approach focuses on the individuality of the patient as a unique person with his or her own unique narrative," Smeets and De Vries note in their contribution. From such approaches to documentation they distinguish nomothetical approaches. These focus more on general, prefabricated classifications. Indicators in such nomothetical approaches can be determined on the one hand through theory and on the other through empirical research. Smeets and De Vries prefer a narrative, idiographic approach to documentation, but see a workable compromise in empirically supported indicators for documentation systems. However, narrative documentation patterns are also problematic in many respects, at least for some authors. Many questions arise: Are the narratives understandable for others? Do they make sense to the members of an interprofessional team? Don't they pose a danger to pastoral confidentiality? After all, don't narrative entries take a lot of time? In addition, do they really benefit the patient? Especially for interprofessional documentation in EMR, a standardized documentation format has many advantages, although it undoubtedly submits itself to a certain institutional logic. On the other hand, the use of standardized formats creates a new opportunity for cooperation in the design of the formats and may facilitate further cooperation, provided that hospitals and healthcare facilities have an open ear. The further integration of spiritual care does not only depend on the spiritual care givers themselves but essentially on the appraisals and judgments made by others, that is, spiritual care providers are dependent on the willingness of other parties to include them in the patient's care. They need medical and other hospital professionals to approve of what they are doing.

3   The Pastor-Patient Relationship and Pastoral Confidentiality

Pastoral confidentiality is a crucial point in many debates on documentation. Repeatedly it has been the subject of discussion. Often the issue is raised by actors who are themselves hostile to documentation. It is important to emphasize that the patient is the bearer of (t)his secret, not the pastor or chaplain. Moreover, there are wide areas of interpersonal communication that do not fall under the scope of pastoral confidentiality. The recording of many encounters is therefore unproblematic. In addition, it is possible to maintain professional secrecy within a team. However, the field is undisputedly delicate; and it is particularly delicate in narrative documentation processes. The case study Smeets and De Vries mention in the main text therefore merits more close examination. The use of documentation in this case may or may not be illegal, but it is certainly morally questionable. For the entry appears to suggest family dissonance. Has the mother of the sick child given her consent to this entry? If not, then I would consider such a narrative entry problematic.

What comes into play in this case – apart from pastoral confidentiality – is the relationship between pastor and patient. In the discussion of his contribution to our workshop, Smeets showed particularly clearly how this relationship can lead to differences in perception. There are divergences. Spiritual care givers often see themselves in a different role than patients do. Intensive empirical research is needed in this area, as Smeets points out. Such research will be easier with spiritual care documentation as a resource. I would like to add that PROMs ("patient reported outcome measures"), that is, feedback created by patients, could be an especially meaningful addition to this field of research. It is undisputed that research work in this area has enormous potential. These considerations really affect the core of patient-oriented care and concern in the hospital context. As many studies have shown in detail, spiritual care givers do not need to have any worries: Their contribution in the clinical context is highly appreciated by patients.

References

Bos, T., A. Kemper, P. Van der Schaft, and A. Van 't Spijker-Niemi. 2003. Een hermeneutisch-diagnostisch model voor geestelijke verzorging. *Tijdschrift Geestelijke Verzorging* 6 (27): 23–39.

Fitchett, G. 2002. *Assessing spiritual needs. A guide for caregivers*. Minnesota: Augsberg.

Lammers, H., and W. Smeets. 2016. De positie van de geestelijk verzorger in juridisch perspectief. In *Handboek spiritualiteit in de palliatieve zorg*, ed. W. Smeets et al., 66–87. Almere: Parthenon.

Smeets, W. 2006. *Spiritual Care in a Hospital Setting. An empirical-theological exploration*. Leiden: Brill.

———. 2009. Gender en lichamelijkheid in de beleving van geestelijk verzorgers. *Tijdschrift Geestelijke Verzorging* 13 (59): 14–27.

Smeets, W., and A. de Vries. 2016. Spiritual screening in a secular context. In *Compassion for one another in the Global Village. Social and cultural approaches to care and counselling*, ed. U. Elsdörfer and T.D. Ito, 92–99. Wien: LIT.

Smeets, W., F.W.J. Gribnau, and J.A. van der Ven. 2011. Quality assurance and spiritual care. *Journal of Empirical Theology* 24 (1): 80–121.

van der Ven, J.A. 1994. Pastorale protocolanalyse II. Pastoraat in maat en getal. *Praktische Theologie* 21 (1): 7–20.

Vandenhoeck, A. 2007. *De meertaligheid van de pastor in de gezondheidszorg. Resultaatgericht pastoraat in dialoog met het narratief-hermeneutisch model van C.V. Gerkin*. Leuven: Katholieke Universiteit.

———. 2009. Op de kaart! Registratie vanuit pastoraal perspectief. *Pastorale Perspectieven* 143: 6–16.

The Spiritual Care Giver as a Bearer of Stories: A Belgian Exploration of the Best Possible Spiritual Care

Anne Vandenhoeck

Belgium consists of three regions, defined by the use of language: the Flemish-, French- and German-speaking regions. In what follows, I will focus on the Flemish part of Belgium as professional spiritual care is more organized and integrated in healthcare there than in the French- and German-speaking regions. This observation is based on the fact that Flanders has the highest number of paid spiritual care givers in healthcare institutions in comparison with the other regions and has a professional association for spiritual care givers.[1] Professional spiritual care in Belgium is mainly provided by Roman Catholic spiritual care givers and Humanist spiritual care givers. Spiritual care givers from those two groups are hired and paid by general hospitals or other healthcare institutions, such as mental health hospitals, nursing homes or homes for people with disabilities. Healthcare institutions are by law obligated to give patients access to the support of a representative of their faith or life view. Representatives of other religions or life views are usually not staff members in hospitals, but external representatives who are called in and are paid per (requested) visit. The profile of a Catholic spiritual care giver in Belgium is no longer that of a male priest. In Flanders, the majority of Catholic spiritual care givers are lay women. The change in staff has also provided a paradigm shift in Catholic spiritual care: since the late eighties it has no longer been focused on sacraments but on narrative hermeneutical pastoral care.[2] Patient and spiritual care giver both enter

---

[1] http://www.pastoralezorg.be/page/beroepsvereniging/ (access 13.04.2019) This association is called: The Professional Association for Catholic Spiritual care givers in Health Care.

[2] See the works of Charles Gerkin, who reinserted a theological foundation in pastoral care after the therapeutic paradigm. His theolgical foundation was based on the hermeneutics of Ricœur and Gadamer, on the theology of Moltmann and Niebuhr. See for example: C.V. Gerkin, The Living Human Document: Re-Visioning Pastoral Counseling in a Hermeneutical Mode, Nashville, TN,

---

A. Vandenhoeck (✉)
Catholic University of Leuven, Leuven, Belgium
e-mail: anna.vandenhoeck@kuleuven.be

© The Author(s) 2020
S. Peng-Keller, D. Neuhold (eds.), *Charting Spiritual Care*,
https://doi.org/10.1007/978-3-030-47070-8_8

129

the encounter with their own horizon of (pre-)understanding. In their meeting, the dialogue between their horizons of understanding can influence them both and lead to a new understanding, adding to the spiritual story of both patient and spiritual care giver.

For some time now there have been spiritual care givers in Flanders who chart in electronic patient files. However, this does not apply to everyone: there are spiritual care givers who have only recently received access to the electronic patient file and there are those who still have no access at all. Thus there is no uniformity regarding charting in spiritual care in Flanders. What is remarkable is the variety of visions that are at the origins of the differences: for one hospital management, it is self-evident that spiritual care givers should have access to patient files and chart, because they contribute to the care of patients. For others, spiritual care givers shouldn't have access to patient files because they are not a recognized medical or paramedical profession. Some managers or directors see spiritual care givers as non-professionals, *einzelgänger* who are there to convert patients to their life views. They have no clue what spiritual care givers do and don't see a reason to give them access to patient files. Spiritual care givers themselves also differ in their motivations of charting. Some are very happy to chart and to have a forum to share their care for patients and loved ones. Others are very hesitant and not convinced of the necessity to chart. Doubts regarding charting almost always have to do with confidentiality.

In this article we limit ourselves to the situation of spiritual care in general hospitals in Flanders. At the request of several Catholic spiritual care givers in general and university hospitals, an exchange group met twice in 2017 about the opportunities and tensions of electronic charting. This includes the perspective of patients, who can view their file online. The text below is partly based on the discussions at these meetings.

1   A Lack of Uniformity Regarding Spiritual Care in Electronic Patient Files

There is a great variety in the possibilities that spiritual care givers have regarding charting. A first determining factor in the differences is the software for the electronic patient files. Some of the developed software programs, especially those of an Anglo-Saxon design, include ample space for spiritual care and are not limited to a box for denomination and/or sacraments received. As a consequence, spiritual care is really integrated in a whole person perspective on the patient. Within this perspective, there is often less limitation in access for spiritual care givers to the charting of other healthcare professionals on the same patient. Even if the patient is transferred or admitted to a different unit than their own, spiritual care givers can obtain

---

Abingdon, 1984. C. V. Gerkin, Widening the Horizons. Pastoral Responses to a Fragmented Society, Philadelphia, PA, Westminster John Knox Press, 1986.

temporary permission to access the electronic patient file, provided that this can be justified. Some hospitals develop their own software for patient files. Examples of this in Flanders are the University Hospitals of Leuven, connected with the Catholic University (KU Leuven). The design of their software provides ample space for spiritual care.[3] This system is shared by all affiliated hospitals. The spiritual care givers of these hospitals formed a task force where they discussed the possibilities and limitations of the software while it was developed and put into practice. In other words: they had an input as spiritual care givers in the development of the system and the language used for their sections. The differences in charting are thus partially down to the differences in software.

But software development is also based on perceptions of healthcare. If the perspective is truly one of the whole person, not merely theoretically but in reality, then spiritual care will also be integrated in charting systems. In an important development, the University Hospitals of Leuven are currently working to develop a new interdisciplinary patient file. The current patient file is still too much based on the kind of charting that doctors and nurses do. Most systems are not designed for paramedical but for medical charting. It is not always easy to be integrated as a paramedic or as a spiritual care giver in a way that does justice to the work and the perspective of a different discipline other than the medical one. The new interdisciplinary electronic patient file will start from the importance of empowering patients, communication, working together and of a whole patient perspective. It is based on WHO's International Classification of Functioning (ICF). Spiritual care givers will have a checklist to assess the spiritual needs of patients and to communicate the outcomes of spiritual care. At the same time, they will have boxes to chart in a narrative way. The checklists for spiritual needs and outcomes are based on the model of *Discipline for Pastoral Care Giving* (Vandecreek and Lucas 2001), which I will discuss later in this contribution.

Spiritual care givers in Flanders try to work as well as they can with the system that is used in their hospital. They testify to the fact that they are looking for ways to make the system work as well as possible in pursuit of the best possible spiritual care. An example of this is the so-called final report within the Nexus EHR system. If a care giver makes a 'final entry' in counselling (or caring for) a patient around the time of his/her discharge, the system considers this care relationship as terminated and the care giver is not informed when the patient is re-admitted. Spiritual care givers who have to work with this system do not make a 'final entry' or report so that they are informed when the patient is re-admitted and they can visit him or her again.

As already mentioned above, a second determining factor in the lack of uniformity is the level of access spiritual care givers have to patient files. The question arises on what principles (legal and others) the hospital managements decide to allow spiritual care givers access to electronic patient files or not. There is no general rule, but it is good to know that all spiritual care givers are hired and paid by hospitals in Flanders and are technically staff members who are involved in patient

---

[3] The software for electronic charting is called KWS: Clinical Work Station.

care. Moreover, quality control systems like JCI have been designed to integrate spiritual care in order to enhance the quality of patient care and facilitate charting by spiritual care givers.[4] Visibility, accountability and transparency are key features in quality control.

A third determining factor in the lack of uniformity is the level of access to the charting of other professional staff in patient files. If spiritual care givers have access to patient files, they do not necessarily have access to the charting of all the other healthcare professionals (and vice versa). There seems to be no uniformity in who has access to whose charting on the patient in the different hospitals in Flanders. In one hospital, spiritual care givers have only access to the charting of the nurses and in another hospital to the charting of nurses and social workers but not, for example, to the charting of psychologists or doctors. Again, the differences in access to charting of other professional groups reflect differing perspectives on team work. A multidisciplinary take will emphasize the importance of different disciplines but not necessarily of working together for the same patient goals as an interdisciplinary approach does. Decisions about access to charting are taken according to the perspective taken.

Among the spiritual care givers themselves, there is no unity either regarding feelings or beliefs about charting. Spiritual care givers who are not allowed to chart in the electronic patient file often feel isolated. This feeling is exacerbated when they see spiritual care moving to the periphery of healthcare. There are spiritual care givers who have the opportunity to chart but deliberately choose not to or write down as little as possible or chart exclusively with document protection so that what they write is not visible to others. Their motivation rests mainly on the value of confidentiality in the spiritual care giver–patient relationship. Other spiritual care givers feel limited in their care or find that they cannot provide optimal spiritual care because they are denied access to parts of the electronic file, such as the reports of the psychologists or the doctors' follow-up papers. Lastly, there are spiritual care givers who chart out of the perception that they are professional care givers like the others and that they have to contribute to team work by means of charting. It is noteworthy that few spiritual care givers have negative feelings towards oral communication about patients with other members of staff. Naturally, charting does not exclude oral communication.

2 Different Forms of Charting Used by Spiritual Care Givers

In the last 10 years, Catholic spiritual care givers in Flanders have regularly committed themselves to reading articles, taking training courses and holding discussions about the challenges and possibilities of charting. A distinction has always

---

[4] JCI stands for Joint Commission International. It is a US-based organization that works internationally towards improving healthcare by evaluating the quality of care provided. See https://www. jointcommission.org (Access 17.04.2019). Other international or European-based organizations aim for the same goals regarding evaluating quality of healthcare.

been between charting for oneself, disciplinary charting and interdisciplinary charting. With charting for themselves, spiritual care givers refer to the notes that they make for themselves about the patient. These notes serve to record the main story lines and details from the patient's story and determine the quality of possible future spiritual care for the same patient. In this regard, spiritual care givers are bearers of stories. Disciplinary charting refers to those notes about the patient that can be useful for colleagues (spiritual care givers) when the patient needs follow-up in the absence of the spiritual care giver or when the patient is moved to another unit where another spiritual care giver works. Interdisciplinary charting refers to sharing information about the patient that can be useful to healthcare professionals from other disciplines in regard to the overall care for the patient. Every form of charting takes place in the interest of the best possible spiritual care for the patient. The answer to the question why these different forms of charting exist next to each other is quite simple: confidentiality. Spiritual care givers in Flanders often struggle with the question how much of what a patient or loved one confides in them can be shared with others.

A second difference refers to charting via a checklist or through a narrative. In a checklist, spiritual care givers tick boxes that say something about the contact with the patient with calibrated terms. This form of charting is most evident in the context of interdisciplinary charting. A second form of charting is narrative charting. Mostly in complete sentences, the spiritual care giver notes several aspects out of contact with the patient that are important for the overall care plan. There are pros and cons for both ways of charting. The experience of spiritual care givers is that checklists are not often read through. On the other hand, it is a much used tool in healthcare. It takes practice and experience to write comprehensive narrative notes without disclosing too much information and breaching confidence. It is not easy. On the other hand, it is more likely to be read.

3 Tensions and Opportunities

Most Flemish spiritual care givers experience the possibilities of charting as positive. The main motivation to chart in an electronic patient file is the contribution that personal, disciplinary and interdisciplinary charting can offer to the patients the best possible spiritual care. Accurate retention of important facts and storylines is a fundamental contribution to high quality and continuous spiritual care. It contributes strongly to the spiritual care giver as a bearer of stories. Colleagues of a spiritual care giver who receive sufficient background information through charting can continue or take over the care for patients during absences. Other healthcare givers who see that the spiritual care giver is involved with a patient will call more quickly when necessary. Furthermore, spiritual care givers also report that they often receive positive feedback from other care givers on what they chart and their charting makes them more visible in an interdisciplinary team. Charting is also perceived as a way of self-care. Spiritual care givers stop between visits to chart and take a break from visiting. They also report that charting is a way to reflect on what they have done, thus enhancing their professional skills. But in addition to the contribution to quality and efficient spiritual care and the positive side effects of better integration, spiritual care givers also experience a number of tensions and disadvantages in charting.

3.1 Protected Files for Charting Spiritual Care

Charting takes time – often, more than spiritual caregivers are willing to invest. That is why they are looking to use the needed time as efficiently as possible. One of the things that turns out to be an obstacle is the difference between charting for oneself and for others. Spiritual care givers are bearers of stories. Patients entrust them with their life stories. The spiritual care giver is expertly trained to see in those life stories the windows that are opened on the spiritual dimension, the meaning system that enables the patient to live as meaningfully as possible, despite the limitations of age, disease or condition. Spiritual care givers carry these stories together with the patient and connect them, if desired, with the greater spiritual stories that are important to the patient. In order to be able to remember the patient's stories, it is necessary for the spiritual care giver to record important facts and story lines. However, these are not relevant for the whole of the care or/and are often told in confidence. In other words, spiritual care givers need to keep more information on a narrative level than they want or can share with other healthcare providers. Nothing is more annoying than meeting a patient again and not remembering what he has entrusted to you during the previous visit a few months ago. That is why many spiritual care givers use the possibility to check a document marked as 'protected' in the electronic patient file. Other healthcare givers cannot open the file. However, there are hospitals that strongly discourage the creation of protected documents because of shared confidentiality in the context of interdisciplinary care or whole person care. They then ask what the value of an electronic patient file if a part of the information is not shared. Does it still show a multi-dimensional image of the patient? But if protected documents are not possible, should the pastor chart separately in notebooks, index cards or word documents outside the electronic file and do double work?

3.2 The Tension Between Being Sent by a Faith Community and Being Hired by a Hospital

The tension between being sent by a faith community and being hired by a hospital is increased by charting in the electronic patient file. It is the tension of being affiliated with a tradition that evokes trust and confidentiality and being a professional care giver paid by the hospital. What can you chart and for whom? These are questions that arise in the context of the trust that the patient places in spiritual care givers. Supporters of charting will state that patients know that every healthcare provider is part of a team and that it is necessary to communicate in order to coordinate the care as well as possible. Good care means shared and coordinated care. Critics will argue that the patient often sees the spiritual care giver as an outsider, someone who does not really belong to the team and whom you can trust. That trust has not only to do with the eccentric place in the care team but also with the background of spiritual care giving. Catholic spiritual care givers suspect that the ecclesiastical tradition of the confessional confidentiality, which is absolute and of the altar as a sanctuary, still plays a role in the perception of the spiritual care giver. In other words, patients have an archaic image of the spiritual care giver as a confidential advisor par excellence. This image is passed on from generation to generation regardless of whether patients are Catholic or not and regardless of secularization. Spiritual care givers notice this image with patients of all ages and life views. Patients are right in their perception that spiritual care givers are part of a faith community, though they are hired by and working in the hospital.

Formally this is confirmed in their contract with the hospital and in their appointment by the faith community to work in the hospital. They are sent by a faith tradition not to convert people but to assist and support them in their suffering. Spiritual care givers feel called or motivated to do so out of their own faith. The trust of patients and patients' perception of spiritual care givers make many spiritual care givers feel uncomfortable with charting. That discomfort or tension has two layers. On the one hand, there is the suspected ignorance of patients around charting by spiritual care givers. They suspect that everyone charts except the spiritual care giver and the cleaning lady or cleaning man. It is no coincidence that both function as persons whom patients trust their life stories with. On the other hand, there is the constant tension about what to chart and what not to chart. Does the patient's trust, enhanced by an archaic layer of confessional confidentiality and a feeling of a free space, allow charting? This tension will increase in the near future when patients can access the charting of spiritual care givers during their stay in the hospital or from their personal computers. At this point in Flanders, patients can already access their patient files, but so far only the medical part. Spiritual care givers think they will chart differently knowing that patients can read what they write. This evolution will change the central focus from charting in order to coordinate care better to charting in order to inform the patient. Spiritual care givers also ask themselves whether patients would not be shocked if they noticed that they are charting about their visits to them. A fundamental question also arises regarding the protected documents. Will patients get access to these?

It remains important to mention that patients do not only see spiritual care givers as persons who keep confidentiality. They also see them as professionals, members of the staff in hospitals. Spiritual care givers wear badges (and sometimes also uniforms), are on call, provide worship services and rituals, are on the payroll, have a code of conduct, etc. It has also been observed that the more spiritual care givers see themselves as professionals hired by the hospital and part of the team (and not so much as being sent by a faith community), the more likely they will chart and have fewer problems with charting, also with confidential issues. Confidentiality is thus also tied to how a spiritual care giver perceives him or herself in the tension between being sent and being hired. Both ends of the spectrum provide a safe space: you never chart because of confidentiality or you chart everything because of your profession. In between the extremes of the spectrum lies the most interesting challenge: discerning how, when and what to chart within the reality of being hired and sent.

4   The Language of Charting

It is remarkable that spiritual care givers, as a professional group in Flanders, do not have just one template to chart with and do not have an agreement on the language they use. This is not just typical for Flanders, but is the case in the rest of the world too. There is not just one system spiritual care givers chart with, but many. Before electronic patient files, each spiritual care giver also had his or her own system or mode of charting. The contemporary practice of charting in different forms (ticking boxes in a checklist or narrative charting and everything in between) and using different words and concepts has the advantage of seeing what works and what doesn't. The downside is that it makes research into charting, the impact of charting and the content of contacts with patients much more complicated.

In what follows I would like to describe how I came to introduce a particular model for spiritual care in Flanders and how the language used in the model still influences charting in patient files today. Around the year 2001, I myself was a spiritual care giver in the University Hospitals in Leuven. The hospitals were being evaluated by the global management consulting firm McKinsey, which was later renamed as Accenture. The goal was clearly to enhance (financial) efficiency. The spiritual care service had to participate in all the exercises the other services had to make. We soon noticed that we did not speak the language of numbers and outcomes. How were we going to survive without losing mandates? I made contact with Larry Vandecreek, a fulltime researcher in healthcare chaplaincy in New York, and asked if he knew of any hospitals or studies where outcomes were being used. He referred me to Art Lucas, head of the spiritual care service in Barnes Jewish Hospital (BJC Health) in Saint Louis, Missouri. The team worked with an outcome-oriented model for pastoral care and had just published about it with Larry Vandecreek (Vandecreek and Lucas 2001). I ended up going to Saint Louis and working with the model for almost a year. From 2003 till 2007 I did my doctoral research on the model and underbuilt it from a theological perspective with the concept of narrative hermeneutical pastoral care (Vandenhoeck 2007). Art Lucas called the model 'The Discipline for Pastoral Care Giving'. In Flanders I introduced the model as the 'Focus Model' as it requires a particular outcome-oriented mind set from a spiritual care giver (Fig. 1).

How would you describe what spiritual care givers do? Art Lucas and his team started from their observations that spiritual care givers assess and act. Up until then, models in pastoral or spiritual care usually focused on the assessment part of the visits to patients and loved ones. The Focus Model also starts with assessment. Spiritual care givers always start with listening to the stories of patients and their

The discipline

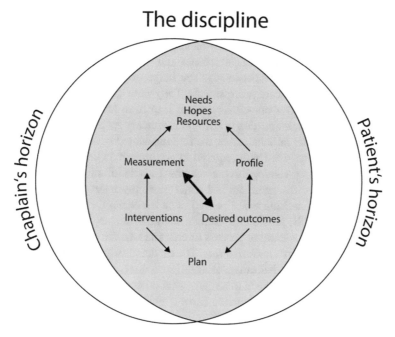

loved ones. They listen with spiritual antennas and are particularly listening to any needs, hopes or resources patients have on a spiritual level. While they listen, they also try to discern how the community around the patient, his/her hope, his/her relation to what is sacred in life and his/her meaning giving functions in relation to being admitted in the hospital. Spiritual care givers, while listening, start to discern how they could make a difference for patients, based on their needs, hopes and resources. The differences your visit makes are outcomes of spiritual care.

An important aspect in the model is that a desired outcome needs to be shared – it needs to be congruent with the overall care plan and it needs to be desired by the patient foremost. In order to make a difference, spiritual care givers have a wide range of interventions that they have built out of their experiences. During the whole visit with the patient, spiritual care givers are focused on the feedback patients give. The verbal and non-verbal measurement of patients determines interventions and outcomes. One of the unique aspects of the model is that it starts from the observation that spiritual needs, hopes and resources (and thus interventions and outcomes) are influenced by medical pathologies. Patients with a chronic lung disease have other spiritual needs, hopes and resources than patients with heart failure. This of course is an incentive for research into shared needs, hopes, resources, interventions and outcomes by medical pathology.

One of the most important assets of the Focus Model is that it gave spiritual care givers a language to communicate within an interdisciplinary context. The language used in the model is a spiritual language. The concept of being a multilingual spiritual care giver was thus introduced. It was Paul Pruyser who stated that chaplains should use a theological instead of a psychological language to assess patients with (Pruyser 1976). But almost 40 years, later a broader spiritual language seems to be more understood by healthcare professionals. The language the Focus Model uses was checked with other healthcare professionals. They were asked if they thought this is what a spiritual care giver does. It is a small leap from language to charting. The model emphasizes communication in an interdisciplinary context. When I was trained to chart with the model in Saint Louis, the training started with checking boxes on a check list. After every visit I was required to go through a checklist and determine what my reason was for visiting this patient, how I would follow up, if there were any recommendations for the team and if I could name my interventions and the outcomes of the visit. Checking boxes proved to be beneficial for learning a language. Going over and over the possibilities introduced me to the language used (Fig. 2).

The next step in the learning process of charting was the challenge to no longer tick boxes but to chart in a narrative way. The model ideally advocates a functional, narrative way of charting. Narrative, as in full sentences, and functional, as in based on the difference between 'need to know' (what other care providers need to know for the best possible care) and 'nice to know' (which refers to sharing unnecessary and often confidential information). Functional, narrative charting is based on five simple principles: (1) How did I end up with this patient? (2) What is my spiritual assessment of this patient? (3) What is the outcome of my contact with the patient? (4) Which interventions did I undertake? (5) What responsibility do I take further for this patient? (accountability). The Focus Model argues for a narrative registration, because experience shows that care providers prefer to read a short piece of text instead of going over a checklist. It also advocates a functional charting that focuses on how contact with the spiritual care giver functions for the patient instead of on the content of the contact.

Two examples of functional, narrative charting:

Initial visit
   Spiritual care giver visited patient x, referral from the attending physician. The patient expressed feeling anxious after a conversation with the doctor about her diagnosis. The spiritual care giver actively listened and explored possible future stories with the patient. The patient felt listened to and expressed feeling calmer. The spiritual care giver will visit the patient again tomorrow.

This charting describes what the spiritual care giver did without disclosing confidential information. The physician referred the patient to the spiritual care giver because she was upset after her diagnosis. When the spiritual care giver visited the patient, she expressed her fear that her husband would leave her after this diagnosis. He had been having trouble coping with her disease and had sought comfort in a relationship outside their marriage. The patient feared that this would definitely break their marriage. Nobody in her family knew of the other relationship her husband had. The patient clearly counted on the confidentiality the spiritual care giver could offer. The charting reveals what is important to the team: the patient felt anxious before the visit and calm and listened to after the visit. What was confidential

Barnes - Jewish Hospital - BJC Health Care   **CHAPLAIN'S NOTE**
                                                   Revised May 23, 2001

Pt. Name_____
Pt. # _____
Room # _____
                                                   Addressograph Area

----

Focus of care: 0 Patient   0 Family Member(s)_____ 0 Both
Date seen: _____ Time seen: _____ am pm   Faith Community: _____
0 Requested by: Name _____ Relationship to patient: _____
0 Initiated by Chaplain   0 Follow-up contact: # _____

REASON FOR VISIT
0 Anxiety   0 Care Path   0 CPR Code   0 Death   0 Family Care   0 Imminent Death
0 Trauma   0 New Diagnosis   0 Pre-surgical   0 Post-surgical 0 Sacraments
0 Support/Counseling   0 Terminal Wean

ASSESSMENT
0 Community Issues   0 Relationship Issues   0 Hope challenges   0 Meaning challenges
0 Coping with Life Change   0 Re-Visioning Future   0 Health Care Ethics   0 Grief/Bereavement
0 End-of-Life Care   0 Beliefs in Conflicts   0 Values clarification   0 Beliefs/Values in Decision-
Making 0 Relationship with Holy   0 Religious Beliefs/Practices   0 Sharing Spiritual Journey

PATIENT/FAMILY CONTRIBUTING OUTCOMES
A. Outcome(s): 0 Anxiety decreased   0 Community identified/engaged   0 Grieving facilitated
        0 Feels connected with The Holy   0 Values clarified   0 Felt heard   0 Hope(s)
        identified/clarified/restored/reconfigured   0 Sense of meaning restored   0 Sense of meaning re-
        visioned   0 Spiritual Resources identified   0 Identified meaning/impact of life changes
        0 Engaged spiritual resources   0 Experience better connection w/tx team   0 Experienced
        supportive presence   0 Ownership in recovery/healing   0 Vented/shared strong feelings
        0 Resources liberated for healing/well-being   0 Experienced relief from expressing feelings
B. Progress: 0 Achieved 0 Partially Achieved/In Process   0 Not Achieved_____

PLAN
A. Continuing Care: 0 No   0 Yes: 0 Daily   0 Every-other-day   0 Weekly   0 As requested
B. Interventions: 0 Crisis Intervention   0 Education Patient   0 Educate Tx Team re
religious/spiritual   0 Grief Facilitation   0 Pastoral Counseling   0 Prayer
0 Sacraments/Ritual   0 Supportive Dialogue

RECOMMENDATIONS (FOR TREATMENT TEAM'S PATIENT/FAMILY CARE)

0 End-of-Life Spiritual Care Template   0 Ethics Consultation   0 Family meeting
0 Interdisciplinary Care Team meeting   0 Initiate DNR Discussion as appropriate   0 Provide time/privacy
for patient/family religious rituals/practices   0 Review Care Plan (diagnosis, condition, treatment,
prognosis) w patient/family

SIGNATURE _____ Pager # _____ 0 Extern   0 Intern   0 Partner
                                                           0 PRN   0 Resident   0 Chaplain
ooooooooooooooooooooooooooooooooooooooooooooooooooooooooooooooooooooooooooooooooooooooooooooo
Date Referral rcvd _____Time rcvd: _____am pm   Length of Contact: _____
Response Category: 0 Emergent (w/in 15 min)   0 Urgent (w/in shift)   0 Routine (w/in day)
Level: 0 Friendship   0 Comfort   0 Confession   0 Teaching   0 Dialogue   0 Counseling
Intensity: (lowest) 0 1   0 2   0 3   0 4   0 5 (highest)

Fig. 2 Checklist. (Adapted with permission from BJC HealthCare. Copyright © 2001 BJC HealthCare. All rights reserved)

thus remains confidential. If the spiritual care giver felt the need to share this information with others because it was relevant for the care of the patient, it would have
been good to ask the patient if this was okay. Another rule in functional, narrative
charting is to stick with the facts and not to give interpretations. Only chart what the

patient did or said: the facts. Use observations and if necessary quotes of the patient to stay as close to the patient as possible.

> *Follow up visit # 3*
> > *Pre-surgery visit.*
> > *Patient expresses feelings about upcoming surgery. The cause for surgery makes patient feel like having to reconceive her future: can she continue to live alone? Spiritual caregiver explored vision of future with patient. To patient 'future' seems to be defined as 'closure' and 'letting go'. Patient asked to be blessed for surgery and upcoming future. Spiritual care giver prayed with patient.*
> > *Spiritual care giver will follow up post-surgery.*

In this short functional, narrative charting, the spiritual care giver indicates some concerns for the whole of the team: the patient doubts her ability to live by herself and she has little hope. The content of the blessing and prayer remains between the patient and the spiritual care giver, but it seemed important to mention that the patient values spiritual support.

Today, the Focus Model and its language still influence the way a lot of spiritual care givers chart electronically in patient files in Flanders. I refer here to the system that was developed within the University Hospitals in Leuven and is used by every affiliated hospital. The categories of outcomes and interventions, of needs, hopes and resources are still prevalent in the charting.

5   Points for Further Reflection

– Charting can form a basis for policymaking and for researchers in spiritual care to carry out statistical operations and to come to an analysis of data. Policies such as care paths or protocols can be changed in order to integrate spiritual care based on charting by spiritual care givers. Charting can show that they are regularly involved in certain patient groups or in certain circumstances. Research based on charting can be executed in order to answer questions such as: Who is involved in the care for the patient? What spiritual interventions take place in visits with patients with certain diagnoses? And, is spiritual care a factor that determines the length of stay? Spiritual care givers' charting can also be used to collect statistical material to strengthen their integration and position in hospitals. Charting can show the value of spiritual care through the interventions or outcomes that are used. Based on charting, the number of mandates for spiritual care givers can be maintained or increased. Charting is therefore an important platform for further research. How can we use it further in this sense? What are the limitations and possibilities?
– The legal aspect is the big unknown and undisputed factor in this case. Who determines who has access to the electronic patient file? Is it legally permissible that spiritual care givers are allowed to chart in an electronic patient file in one hospital and not in another? Which professions are included in charting and which are not? Is there a legal basis for this, or do hospitals just have their own say? Who has access to whose entries? What needs to be written and what does

not need to be written? Are there limits to confidentiality and what is shared confidentiality?

- The view of a patient on the spiritual care giver and the view of a professional healthcare context on the spiritual care giver can differ. For the patient, a low professional profile (the outsider who is a symbol of confidentiality and who is just present, has time and listens) for the spiritual care giver seems attractive. For other care givers and management, the opposite is often the case. They want to see a highly professional spiritual care giver who contributes to the whole of care and can communicate about his or her contribution, preferably in terms of interventions and outcomes. What does that mean for charting? What does that mean for the patient's access to the charting of spiritual care givers? It seems to be another tension where the spiritual care giver needs to balance between two extremes: non-professional and professional.

- A lot of the discussions on charting are in need of thorough theological reflection. For example: the language that is used for charting could benefit from a dialogue with theology. What language should we give to patient with spiritual issues and their loved ones to help them deal with their hospital stay? Which language can be understood by patients and healthcare professionals? Can spiritual care givers use a theological language? Or is their theological perspective a private one, the mother tongue they use among themselves but not in communication with others? And what is the value and goal of a theological reflection? What difference can the expertise of theologians make for the practice of spiritual care and charting?

6   Recommendations

- The most important benefit of charting for spiritual care givers is its contribution to whole person care and to the provision of the best possible spiritual care. Charting serves the best possible interdisciplinary and spiritual care. Therefore, the focus can never be on charting alone. Charting serves spiritual care and not the other way around. This discussion needs to be held on the level of paradigms and values in healthcare regarding whole person care.

- The tensions a spiritual care giver experiences regarding charting are not necessarily negative. They challenge spiritual care givers to find a balance in between extremes and serve as a constant reminder to improve the practice of spiritual care. The tension regarding confidentiality, for example, can function as a constant reminder to chart as if the patient is reading or is present.

- It is important to keep communicating with other professions in healthcare about charting and how it can serve the best possible care for patients. Other caregivers feel the same inner conflicts or doubts about charting confidential information trusted to them by patients or loved ones. They can be partners in that regard to reflect with managers and IT staff about charting forms and possibilities. Many caregivers who are not nurses or doctors do not feel that the models of charting in electronic patient files are designed for them.

Commentary

Eva-Maria Faber (✉)
Theological Faculty of Chur, Chur, Switzerland
e-mail: eva-maria.faber@thchur.ch

1 The Best Possible Spiritual Care

It's like a refrain echoing throughout her whole article: Anne Vandenhoeck explores the *best possible spiritual care* as the primary goal of spiritual care givers. The best possible spiritual care involves helping and supporting patients in their suffering; it implies encounters where patients can express their needs, their sorrow and their hope and where they can entrust their stories to spiritual care givers that are willing to serve as bearers of these stories.

Anne Vandenhoeck rightly points out that charting is not merely another element in addition to these tasks of spiritual care givers but rather a means for better administering the broad spectrum of possible interventions.

The facilitating role of charting is particularly clear in connection with the task of bearing stories. Bearing stories involves remembering them – even when patients are discharged and later reappear. Charting promotes continuity in the provision of spiritual care.

Charting implies reflection that will improve future interventions. By documenting encounters with patients and by revisiting the documentation of previous encounters, spiritual care givers may discover hidden messages in what has been said and improve their understanding of how best to provide support in the future. Vandenhoeck also mentions a restorative function for spiritual care givers themselves: Charting allows carers to take a break and prevents them from carrying the burden and shadows of one patient to the next.

It is worth emphasizing this positive effect of charting on the provision of spiritual care to patients. Although the introduction of spiritual care charting was in response to the administrative need to monitor the interaction of the chaplaincy with other professions, the task of charting has in fact raised the quality of spiritual care given. Patients also stand to benefit from the greater inclusion of chaplains within interprofessional care-giving teams. For the increased visibility of the chaplaincy brings with it increased attention to the outcomes of spiritual care.

2 Contributing to a Holistic Approach

A crucial point concerns the question how far spiritual care givers can and should open their charting to interdisciplinary exchange. With regard to the different forms mentioned (narrative charting, ticking boxes in checklists), it seems reasonable to reserve narrative charts for personal use (protected files that help one to remember details of stories that, however, should not be shared with others or, at most, with fellow spiritual care givers). The checklists could serve as a way to give a more sparing account of the engagement of spiritual care givers.

However, the restriction of narrative charts for personal use would mean that interdisciplinary charting would be an instrument to strengthen the chaplaincy's position rather than enhance holistic care for individual patients in a hospital.

In discussions about her approach, Anne Vandenhoeck established that chaplains face a challenge concerning multilingual competence: The checklist is a tool that helps chaplains to speak a language that is understood by all healthcare givers. But at the same time it tends to focus on outcomes that are easily expressed in economic terms. For spiritual care givers, however, only narrative charting can assist them in bearing patients' stories. It is therefore important for them to participate in the development of the systems in order to obtain a form of documentation that corresponds to spiritual requirements. That means: only if spiritual care givers are allowed to and are willing to contribute their distinctive perspective on the patient can they take their responsibility for a holistic approach in healthcare. Only then will they promote the best possible interdisciplinary holistic care for patients.

In fact, Anne Vandenhoeck observes that, even in interdisciplinary contexts, narrative charting is more likely to be read than other forms of charting. This means: stories are a language that is also understood and appreciated from the perspective of medical agents.

3 Confidentiality

This is where the question of confidentiality arises that haunts the discussions about interdisciplinary charting of spiritual care. Vandenhoeck provides helpful distinctions concerning this important issue. She points to the distinction between facts and interpretation and furnishes different descriptions of the same situation: one that includes details of a patient's worries; the other to be shared without disclosing confidential aspects of his or her anxieties. Another way to meet standards of confidentiality is to focus on functional outcomes for the patient and not on possibly confidential communications. Charting will then deal more with the interventions that are intended to make a difference for patients (an inspiring wording of Vandenhoeck!) and not with information about the patient's story.

In addition to these very helpful indications it could be stimulating to reflect on the topic of the stories. If spiritual care givers are – in Vandenhoeck's beautiful description – bearers of stories, the question "What can you chart and for whom?" can be reformulated as: "What are the stories entrusted to spiritual care givers and for whom?"

Often stories are in fact entrusted confidentially to spiritual care givers. Patients may explain, or sometimes rather tentatively vocalize, how they experience their story – one that may have taken an unexpected turn as a result of a serious incident. They express their inner turmoil, their emotional perspectives on their situation, their shame. These stories are articulated in the presence of a spiritual care giver whose listening may help them to cope with a new situation. The spiritual care giver is a bearer of these stories not for others but rather a co-bearer of the story in solidarity with a person who – in telling a story – is trying to find his or her way in life. However, there may be aspects of a story that transcend the privacy and intimacy of a confident encounter. Nobody is *only* a patient. Whereas the medical history records the progression of an illness, the spiritual history commemorates the story and stories a person has lived and is living. Thus, spiritual care givers will often be bearers of stories not only by listening but also by reminding patients of their own stories when they become frail. Moreover, by charting some of the "public" aspects of these stories, spiritual care givers bear the stories to other members of the healthcare staff and help them to see the patient in the hospital as a whole human being.

4 Discernment

The best possible spiritual care is not a clearly defined measuring unit. It is a search item that depends on persons and situations and stories and, last but not least, on the feedback that patients give. This insight calls for responsibility on the part of spiritual care givers. They are called to discernment. For this reason, Anne Vandenhoeck even finds value in the tensions that spiritual care givers experience, for example in being part of the hospital team as well as belonging to the tradition of a faith community (or being sent by it). Tensions help one stay attentive – in pursuit of the best possible spiritual care.

References

Gerkin, Charles V. 1984. *The Living human document: Re-visioning pastoral counseling in a her-meneutical mode*. Nashville: Abingdon.

———. 1986. *Widening the horizons. Pastoral responses to a fragmented society*. Philadelphia: Westminster John Knox Press.

Pruyser, Paul W. 1976. *The minister as diagnostician. Personal problems in pastoral perspective*. Louisville: Westminster John Knox Press.

Vandecreek, Larry, and Arthur M. Lucas, eds. 2001. *The discipline for pastoral care giving: Foundations for outcome oriented chaplaincy*. London: Routledge.

Vandenhoeck A., Depoortere Kristiaan (sup.) 2007. *De meertaligheid van de pastor in de gezond-heidszorg. Resultaatgericht pastoraat in dialoog met het narratief-hermeneutisch model van C.V. Gerkin*. (thesis) LXX,. 334 p.

Charting in Switzerland: Developments and Perspectives

Pascal Mösli

Pastoral documentation is developing rapidly in Switzerland: it is now an established part of the palliative complex treatment required of hospitals by guidelines drawn up by pastoral teams in the interprofessional context of mainly large hospitals and endorsed by pastoral expert committees. Despite this development, it remains controversial whether pastoral documentation is appropriate. In order to understand how pastoral care professionals throughout Switzerland think about documentation, a survey of German-speaking pastoral care professionals was conducted in Spring 2019.

This article gives an overview of the most important developments in pastoral documentation, summarizes debate surrounding it, and presents the results of the survey among chaplains. The documentation practice of three hospitals is then presented in order to provide a deeper insight into the concrete practice of documentation before the final section discusses some of the questions and future challenges pastoral documentation in Switzerland faces.

1 Introduction[1]

Healthcare chaplaincy in Switzerland operates in a very diverse regulatory landscape. Depending on the canton, there are different legal and ecclesiastical framework conditions. In many cantons, hospital chaplains are church employees; in

---

[1] I would like to thank Simon Peng-Keller, who supports me a great deal in my research work and with the writing of this article, as well as my colleagues Saara Folini-Kaipainen, Claudia Graf, Livia Wey-Meier, and David Neuhold from the National Science Foundation research project Documentation, with whom I am in a creative, exciting dialogue. They have all read the article critically and given me a lot of helpful feedback.

P. Mösli (✉)
University of Zurich, Zurich, Switzerland and Reformed Church of Bern
e-mail: contact@pascalmoesli.ch

others, they are employees of the relevant institution; and in some cantons, the state contributes funds. In contrast, the professional prerequisites for pastoral work in hospitals are largely uniform: a university degree in theology, additional pastoral psychological training, and a mission from one of the three churches that are officially recognized in most Swiss cantons. These churches are the Roman Catholic Church, the Reformed Church, and the Christian Catholic Church. (The Jewish community is also recognized in many cantons.) 60.9% of the Swiss population are a member of one of these three national churches (36.5% Roman Catholic and 24.4% Reformed, with around 15,000 members of the Christian Catholic Church), 5.8% belong to other Christian communities, 5.2% are Muslim, 1.7% belong to other religious communities, and 25% are nondenominational.[2]

There are two national umbrella organizations for German-speaking healthcare chaplaincy in Switzerland: the Reformed and the Roman Catholic pastoral care associations. In some cantons, but by no means all, there are professional standards for pastoral care in hospitals, but there are none at the national level.

Until 5 years ago, most chaplains in Switzerland did not contribute to interprofessional charts. Most chaplains would not have considered doing so for a moment, and those that did would most likely have rejected the idea for reasons of confidentiality. On the other hand, many chaplains have been documenting their work for their own purposes for a long time. For example, they may have recorded which topics were discussed in a conversation and use the documentation as an aide-memoire, to help maintain the relationship with the patient and for reflection.

Another field in which a form of documentation is used is training: For many years, excerpts of conversations have been used in Clinical Pastoral Training (CPT) courses for educational purposes. The purpose of the documentation here is therefore to help one reflect on one's own professional actions. Each week, the participants in a course bring with them minutes from a meeting in their field of practice. There is also a manual for writing the minutes of a discussion. The verbatim reports are analyzed and discussed in the training group, sometimes used as a template for a case discussion and then returned to the authors. This is an established and effective learning method in the context of pastoral training.

Only in a few hospitals, usually larger ones, have chaplains been charting in an interprofessional context, or within the pastoral care team, for more than a few years. Often this has taken place where pastoral care is administered on behalf of the institution or the state and is also financed by the latter. I will come back to this point.

---

[2] Figures for the year 2017, Federal Statistical Office: https://www.bfs.admin.ch/bfs/de/home/statistiken/bevoelkerung/sprachen-religionen/religionen.html, accessed 20 July 2019.

2   The Current Development

The situation in Switzerland has been in a state of flux for about 5 years now. This change has to do with the digitalization of the healthcare system and with the implementation of electronic documentation methods. The electronic patient file (EPD) is a collection of personal documents with information about the health of the patient. This information can be accessed via an Internet connection. Patients grant access to caregivers and determine who can view which documents. The law and implementing provisions on electronic patient files came into force on 15 April 2017. Since then, hospitals have had 3 years to introduce the electronic patient file, while nursing homes and birth centers have had 5 years. This rapid development in the healthcare system raises the question of whether pastoral care should feature in interprofessional, electronic documentation.

But the mentioned change also has to do with developments in professional pastoral care itself – especially in the fields of emergency, palliative and spiritual care, and relating therefore to the concerns of interprofessional cooperation. These developments are described below.

2.1   Emergency Care

An important context in which documentation systems have developed is emergency pastoral care. On the one hand, the documentation serves as an information basis for intraprofessional work to enable other chaplains in the team to continue the work of companionship. On the other hand, documentation can demonstrate to the institution the importance of this care provision and the corresponding staff resource requirements.

2.2   Palliative Care

The decisive step for the integration of chaplaincy as part of the palliative care teams was taken by the National Strategy for Palliative Care of the Federal Office of Public Health,[3] which regards spiritual care as an integral part of palliative care (following WHO standards). Interprofessional cooperation including pastoral care thus became standard for palliative care. This development was further supported by the financial incentive system. Since 2011, palliative services in Switzerland have been billed within the framework of the CHOP (Swiss Operation Classification[4]) system.

---

[3] Bundesamt für Gesundheit (BAG).

[4] Version 2019: www.bfs.admin.ch/bfs/de/home/statistiken/kataloge-datenbanken/publikationen.assetdetail.5808569.html

Pastoral caregivers and other psychosocial professional groups play an important role in fulfilling the two CHOP codes 93.8A.2 and 93.8A.3. These stipulate that the multidisciplinary treatment team, consisting of a doctor, nursing staff, and staff in at least two of the therapeutic areas (social work/pedagogy, psychology, psychotherapy, ergotherapy, speech therapy, nutritional counseling/therapy, pastoral care, art therapy), will work a total of at least 6 hours per treatment week. In many hospitals in Switzerland, specialists invoice their services electronically in a form specially created for this purpose, which automatically evaluates the criteria for the complex code. Support within the framework of palliative complex treatment is the only pastoral service in Switzerland which has direct financial consequences or which has an impact on a hospital's income. It belongs to the group of services that must be fulfilled in order to receive the financial contributions of the insurance companies and to guarantee the (re)certification of the palliative wards. The prerequisite for this, however, is the documentation of pastoral services, if only in respect of the duration of pastoral attendance. If the chaplain is not prepared to keep records, the ward will (have to) use other services, like the psychological service. Pastoral care thus runs the risk of dropping out of the treatment team.

2.3   Spiritual Care

In recent years, we have seen significant and quite diverse developments in spiritual care as an interprofessional task in Switzerland – in the field of palliative care and beyond. In this context, two documents have emerged that are very important for our topic.

First is the document of the "Spiritual Care Taskforce" of "palliative.ch", the Swiss Society for Palliative Care[5]: "*Spiritual Care in Palliative Care. Guidelines on Interprofessional Practice.*"[6] This document contains the following statement on documentation: "Interprofessional Spiritual Care requires intensive communication, including medical documentation as an important tool. Requirements for the documentation of R/S[7]: (a) it is part of the documentation system accessible to all members of the treatment team; (b) all professional groups involved are requested to chart observations and agreements on R/S if they are relevant to the treatment and (c) the special and context-specific professional framework conditions regarding the possibility and limits of pastoral communication must be taken into account (keyword: pastoral secret)" (Spiritual Care 2018, 12f).

Second, in June 2019, the professional pastoral care group of palliative.ch published its guidelines "*Pastoral Care as Specialized Spiritual Care in Palliative*

---

[5] The Spiritual Care Taskforce supports interdisciplinary work toward a better understanding of Spiritual Care in Palliative Care. It is made up of experts from various professions.

[6] The guidelines ("Spiritual Care in Palliative Care. Leitlinien zur interprofessionellen Praxis") can be downloaded here: https://www.palliative.ch/de/fachbereich/task-forces/spiritual-care.

[7] R/S means religious/spiritual.

Care".[8] Guideline 7 (access to information and documentation of care) broaches the issue of documentation: "As a member of the treatment team, the pastoral care specialists have access to information that will allow them to adapt their concept of spiritual-religious support. [...] The formal and/or informal exchange of information contributes to the strengthening of the team spirit and a positive culture within the treatment team. The information is recorded in the part of the electronic data record intended for pastoral care. In the electronic patient file spiritual/religious elements can be noted; the elements of the electronic patient file are available for all involved experts; the pastoral care specialist is in contact with the interprofessional team" (Leitlinien 2019, 14f).

2.4   Conclusion

The developments described above show that the topic of documentation within pastoral care has developed intensively in recent years and gained in importance. External factors, such as institutional requirements and the financial settlement of complex palliative services, have been crucial, but so also have developments within the profession of pastoral care itself. The insight that documentation could contribute to the professionalization of pastoral care was increasingly important. Alongside the initiatives of some pastoral care specialists, important roles were played by pastoral care units in large hospitals and professional associations, such as the Protestant and Roman Catholic National Pastoral Care Association and the Professional Pastoral Care Group of palliative.ch. The National Pastoral Care Associations have recognized the importance of the topic since 2013. They initiated the research project for documentation and organized the first conference on the topic. In order to better understand the importance of documentation for chaplains and their attitude toward it, a survey with the support of the Pastoral Care Associations was commissioned. This survey will be presented in the next two sections.

3   A National Survey

3.1   Occasion

It was remarkable that all discussions about documentation – be they conferences or collegial meetings – provided an opportunity to discuss fundamental topics of pastoral work and its profile. When confronted with the documentation topic, chaplains discussed the possibilities and limits of language; the connections between spiritual

---

[8] The guidelines ("Leitlinien. Seelsorge als spezialisierte Spiritual Care in Palliative Care") can be downloaded here: https://www.palliative.ch/de/fachbereich/fachgruppen/fachgruppe-seelsorge.

and religious matters with physical, psychological, and social phenomena; the question of individuality; and so on. Also interesting was the high emotional intensity with which the documentation was discussed. The topic of documentation hardly ever leaves anyone cold. Obviously, the topic challenges pastoral professionals to take a stand.

But what exactly do chaplains working in institutions think about pastoral documentation? How do they assess documentation in the context of the development described above? Do they chart themselves? Would they be prepared to do so? – Or is documentation not an appropriate pastoral instrument, in their view? These and other questions were put to chaplains in order to better understand their self-conception and practice. The results of this survey are explained and discussed below.

3.2   The Participants

In March 2019 we sent the questionnaire to 142 Protestant and 134 Roman Catholic chaplains who are members of the Reformed and Roman Catholic Pastoral Associations. Most institutional chaplains in Switzerland are members of one of these two national associations.[9] In total there were 276 people, 5 of whom were retired. Therefore, 271 people were relevant for the survey.

One hundred and forty eight chaplains completed the questionnaire, which corresponds to a response rate of 54%. We can therefore say that the survey provides a good insight into the Swiss situation.

Among the participants, 77 were Protestant chaplains, 70 Roman-Catholic chaplains, and one chaplain belongs to another denomination (unknown to us). 104 of the participants work in a hospital, 21 in a psychiatric clinic, 37 in old-age institutions, and 20 in other contexts (Fig. 1).

An important factor for questions about charting is the appointing authority. The employment conditions of institutional pastoral workers in Switzerland are complex: some employers are churches; sometimes the state plays this role and sometimes the institution. Moreover, there are also mixed forms of employment.

Thirty one participants in our survey are employed by the institution they work in, 120 by the church, and 16 by the government. One person has a different employer (Fig. 2).

The different employment arrangements bring with them different expectations, with consequences for the perception and assessment of documentation:

– With regard to health service providers, pastoral care must make clear what it offers as its service, which quality standards it is committed to, how it secures these, and what framework conditions and resources it requires for this purpose.

---

[9] Efforts are underway to establish the number of institutional chaplains in Switzerland. However, no results are available yet.

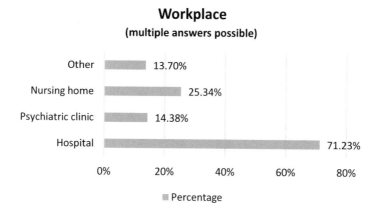

Fig. 1 Workplace

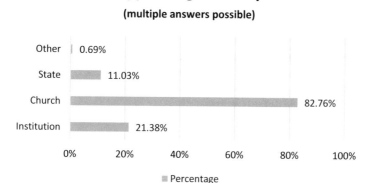

Fig. 2 Appointing authority

- With regard to the institutions of the health system, pastoral caregivers must show what specific contribution they make in interdisciplinary cooperation and how they network with other occupational groups.
- With regard to the state and social authorities, pastoral caregivers must show that they do not only care for church members but also for other members of the multicultural and multi-religious environment. They support people of all worldviews in a non-exhaustive way. They are thus oriented toward an open society that stands up for human rights and human dignity.
- With regard to the churches, pastoral caregivers must show how they justify their mission from within their religious tradition and how they fulfill it in a contemporary way.

The employment arrangements of chaplains are an important framework condition for the questions of the necessity of interprofessional cooperation and of the depth and nature of the integration of pastoral care in the institution. Such arrangements therefore also have an influence on the treatment of professional and pastoral secrecy, which will be discussed later.

4   The Results of the Survey

4.1   General Results

An important result is simply the response rate: Documentation is a topic that many chaplains are occupied with or want to deal with. This finding is confirmed by the many responses from participants (more than 50% of those who took part in the survey), who told us that they were willing to participate in further discussion.[10]

It is also clear that the confessional background of chaplains does not influence response rate: Of the Protestant chaplains, 77 participated, and of the Roman Catholic chaplains, 70. In relation to the questionnaires sent, this corresponds to a rate of 55% of Reformed chaplains and 53% of Catholic chaplains.

It can also be assumed that the topic of documentation is more important in hospitals than in other institutions. 57% of the data were received on the institutional field of hospitals. This is perhaps a reflection of the interprofessional integration of pastoral care, which is tightest in hospitals.

4.2   Charting: Yes or No?

But to what extent do chaplains in Switzerland record institutional pastoral care in interprofessional documentation? All participants answered this question. 44 of the participants contribute to interprofessional documentation; 27 chart only in the context of complex treatment in the field of palliative care; 74 chaplains do not chart. In summary, 49% of all chaplains contribute to documentation in an interprofessional context (including palliative care); 51% do not chart at all. We will break down the number of those who currently do not chart further in connection with the question of who can imagine charting and who cannot (Fig. 3).

However, the fact that a chaplain does not currently chart in an interprofessional context does not necessarily mean that he or she does *not want to* chart now or in the future. There are several reasons why chaplains cannot chart interprofessionally or are not allowed to. The institution, as well as the commissioning church, may prohibit this, and there may be no possibility of making pastoral entries in the existing

---

[10] The field at the end of the survey was "contact permission."

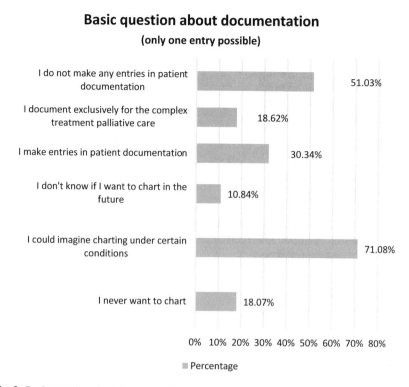

Fig. 3 Basic question about documentation

documentation. We therefore asked those who are not currently charting whether they would prefer not to chart under any circumstances or whether there are circumstances under which they would like to chart. Out of a total of 83 answers to this question, 15 people stated that they never wanted to chart and 9 people said that they do not know whether they want to chart under different circumstances. 59 people stated that they would chart under certain circumstances – 71% of respondents.

While the question above was about interprofessional documentation, we also asked chaplains whether they contribute to documentation for themselves (with nobody else having access) and whether they contribute to documentation for the members of the team (with access restricted to the members of the pastoral team). 109 respondents answered this question. 96 people chart for themselves and 39 for the members of the team. With this question, it was possible to agree to both answers. We therefore do not know whether there are chaplains who chart only for members of the team but do not keep additional records exclusively for themselves (Fig. 4).

However, we can see that 88% of all participants chart for themselves, while only 49% contribute to interprofessional documentation. We can conclude that documentation for one's own work is part of pastoral workers' common practice, while interprofessional documentation is on the rise. It would be exciting to investigate whether

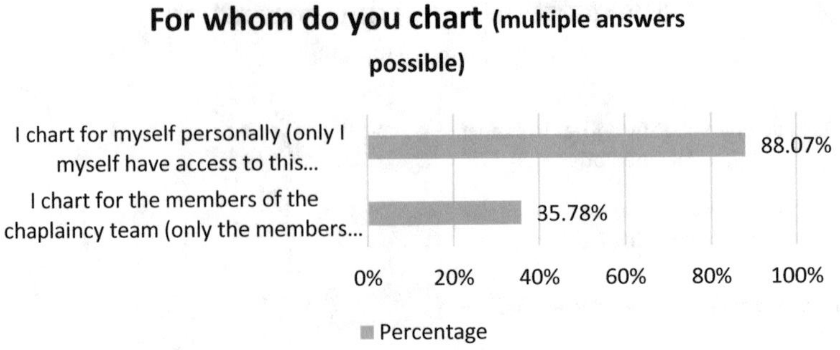

For whom do you chart (multiple answers
possible)

Fig. 4   For whom do you chart

personal documentation follows a conceptual basis or whether it is a rather acciden-tal documentation of important experiences or insights, in order to be able to further inform the accompanying process.

4.3   Reasons to Document

We have already seen that employment arrangements, and the context of justifica-tion they create, are significant determinants of documentation practice. Pastoral workers were asked why they do or do not chart, and 132 responded (16 people skipped the question).

Most chaplains contribute to documentation on the instructions of their institu-tion (65 people, 49%), while almost as many chart for personal reasons (61 people, 46%). 39 people (30%) chart in accordance with a team decision, 22 people (17%) in accordance with the decision of a superior, while 8 people (6%) do so on the instruction of their church (Fig. 5).

Given the initial situation outlined above, the findings are not surprising. Healthcare institutions have an interest in somehow being able to measure the effects of professional interventions, and in the case of the palliative complex treat-ment already mentioned, such measurements have financial significance. Where pastoral care work is on behalf of the institution and is paid for by it, it is naturally perceived to be part of the work of the institution, which often leads to increased interprofessional cooperation. In the context of this, documentation can be an important instrument of communication.

The interprofessional integration of pastoral care has already been discussed sev-eral times. When asked to document their reasons for charting, 24 out of a total of 36 comments in the free text fields referred to interprofessional cooperation. This is regarded as an essential prerequisite for the practice of pastoral care. Interprofessional cooperation was also cited in most of the comments in the free text field about the circumstances under which chaplains might consider charting.

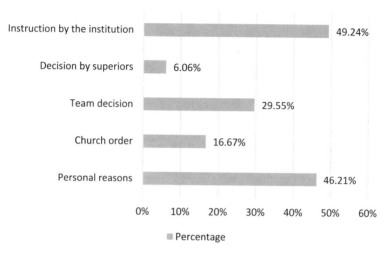

I am charting for the following reasons / I am not charting for the following reasons

(multiple answers possible)

Fig. 5 Reasons for (not) charting

4.4   *Confidentiality*

Professional secrecy is a constraint on the exchange of information. Determining the nature and extent of the constraint pastoral secrecy imposes relationship is in itself a challenge for pastoral work, especially in the context of interdisciplinary cooperation. One response to this challenge is a handbook published in 2016 by the Federation of Swiss Protestant Churches on the subject (Famos et al. 2016). The various Reformed cantonal churches in Switzerland, as well as the Catholic dioceses, are familiar with different forms of the pastoral confidentiality. There are different views as to what, if anything, of the substance of an encounter it is permissible to pass on in interprofessional discourse and what scope there is for interprofessional cooperation and what practices are prohibited. The question of how pastoral secrecy relates to professional secrecy is also controversial.

The pastoral secret was not an explicit topic of the survey, but it did play a role in the respondents' answers. Chaplains who do not contribute to interprofessional documentation were able to explain their reasons in a free text field. 56 people took up this opportunity (42% of all entries). Of these 56 persons, 8 cited the pastoral confidentiality as a reason why they do not document (12 mentioned that they use other means of communication, 27 said that their institutional context does not allow them to document, 9 gave further reasons). It must be stressed that we did not explicitly ask about the importance of the pastoral secret, so the number of responses should be interpreted with caution. Taken together with the other results, one could perhaps give the following summary: the pastoral secret plays an important role in the whole discourse, but it is not principally a stop signal for the possibility of inter-professional documentation.

4.5   The Practice of Documentation

Finally, we asked the chaplains about their charting practices. The responses were as follows:

- I chart regularly: 40 replies.
- I chart if necessary: 36 replies.
- I chart according to specific guidelines: 28 replies.
- I chart not according to specific guidelines: 12 replies.
- I chart by hand: 17 replies.
- I chart electronically: 54 replies (Fig. 6).

What is noticeable is that only a few chaplains chart according to specific rules or guidelines. This reflects the earlier observation that documentation as a standardized instrument is still in its infancy.

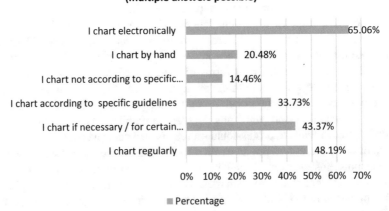

Fig. 6 Information in case of interprofessional charting

5   Three Examples of Charting Systems in Switzerland

Now that the survey has provided a detailed picture of the attitudes and practices of chaplains in institutions in Switzerland, this section will outline a more concrete picture of existing documentation models in Switzerland. There is a wide variety of documentation practices in Switzerland. In some cases, documentation extends only to registration; in others, the involvement of pastoral care is registered without any further information; and in some cases, pastoral information is recorded to inform assessment-based treatment.

In his article "How do we measure quality in charting?" David Lichter (2015, 3ff.) described three forms of quality assurance in charting:

- Process measures focus on processes that lead to a certain outcome.
- Outcome measures focus on results.
- Performance measures relative to professional standards focus on the measurable standards agreed upon across the profession.

As shown above, professional standards for pastoral documentation have been developed in Switzerland in the context of specialized palliative care. These standards would be the basis for the form of quality assurance mentioned under (c). The consistent implementation of these standards, as well as the other forms of quality assurance mentioned by Lichter, is still a long way off in Switzerland. The standards express the ambitions of chaplains working in palliative care have for documentation, rather than requirements implemented in practice.

The following three examples come from pastoral care units in large hospitals in the cities of Zurich, Bern, and Lausanne. The documentation models in Zurich and Bern, like most Swiss models, have been developed as a pragmatic response to the need to deepen and consolidate interprofessional cooperation. The model in Lausanne is an exception to the Swiss norm, because, on the one hand, it has a pastoral-conceptual basis and can therefore be linked to the first two forms of quality assurance of process measures and outcome measures; and, on the other hand, it has been conceived as an instrument for interprofessional cooperation.

5.1   Triemli Hospital in Zurich

The "Stadtspital Triemli" is a central hospital in the city of Zurich with a supraregional catchment area. With 467 beds and around 3,000 employees, it provides around 400,000 people with comprehensive basic and emergency care. The pastoral care team consists of seven pastoral workers (three Reformed and four Catholic).

Pastoral care has gained access to the electronic documentation system in the hospital, but it has been a long road to this point. Only after several requests were the hospital's responsible persons prepared to set up a working group to deal with this issue. This working group was set up to determine what information from nursing documentation it is necessary and important for pastoral care workers to be able to access if they are to participate in the framework of interprofessional cooperation. The vessels of the electronic patient documentation, into which the pastoral care information can be fed, were then defined, so that the treatment team, in particular the nursing staff, has access to it. The results of the working group were submitted to the management of the hospital for appraisal and received approval. Since then, the pastoral care department has had partial access to the electronic patient documentation of all patients hospital-wide, passively (i.e., information can be collected) as well as actively (i.e., information can be fed in).

This access to the information system WiCare is restricted to selected modules, such as patient information, reports, and protocols as well as the Departure/ Transfer field.

Pastoral care information is fed into the system and may record, for example:

- That the patient was visited by a chaplain or is visited regularly
- That faith, religion, or spirituality is an important resource for the patient
- That the chaplain should be informed about a deterioration of the general health or about the presence of relatives
- That contacts have been established with pastoral workers of other languages, denominations, or religions
- That an anointing of the sick has taken place or is planned

As the healthcare chaplains themselves note, this information about the patients does not have a great direct influence on the content of the pastoral work, but it is helpful for pastoral organization. Access to WiCare simplifies the interprofessional flow of information and makes it more reliable. It raises the profile of pastoral care with the treatment team as the work of the pastoral care team becomes more visible. This in turn leads to greater interprofessional networking. Information flows more directly between the people involved (nurses, doctors, therapeutic services, pastoral care, etc.). Less information is lost, and stored information can be quickly viewed by all. This leads to less duplication and thus helps to ensure good patient care. "Our entries in the system make our work visible to all carers. We show that we are part of the interprofessional care team and that we can often act as a hinge between patients, relatives and caregivers."[11]

---

[11] Martin Rotzler (catholic leader of the pastoral care team) and Christoph Wettstein (chaplain). This whole section about the pastoral documentation in the city hospital Triemli owes much to their information and assessments. I'm very grateful for their support.

5.2   Inselspital, University Hospital in Bern[12]

The Inselspital employs around 8300 employees who look after about 44,000 inpatients and over 300,000 outpatients. The ecumenical pastoral team consists of 10 Reformed and Roman Catholic chaplains. The team positions itself as a component of holistic care and is, as such, a supportive service alongside others (e.g., social counseling, psycho-oncology, and palliative care) within the extended treatment team. The pastoral team is in contact with the nursing staff and the ethics department and participates in interdisciplinary (IDR) and interprofessional (IPR) reports. It obtains information from the nursing anamnesis and the respective screening or assessment models of the clinics – and its work can be integrated into the therapy plans (e.g., in neurology and neurorehabilitation).

5.2.1   Interprofessional Documentation: i-pdos

The i-pdos contains a hospital-wide IT-supported patient file to which various occupational groups have access. Medical, nursing, and therapeutic documentation have central importance. Access rights are regulated individually. The "I" in the program name means "integrated" but also "Inselspital-wide," "IT-aided," "interdisciplinary," and "interprofessional." Pastoral care in i-pdos does not document the content of conversations but only the reason for the meeting, the mode in which it was conducted, and arrangements for any future pastoral care. The pastoral entries in the i-pdos inform other professional groups about pastoral intervention and serve as a basis for in-depth, interprofessional cooperation (Fig. 7).

5.2.2   Internal Documentation: Seel:is

In 2012, after other pastoral and psychological assessment tools had been examined, the internal pastoral information system "Seel:is" was introduced, following an intensive development and evaluation phase within the pastoral team after intensive study of other pastoral and psychological assessment tools. The entries in the Seel:is are, in contrast to those in the i-pdos, only visible to the pastoral team members and serve the purposes of intraprofessional cooperation as well as quality assurance for individual chaplains and the pastoral team as a whole. The system essentially consists of two tools, the cockpit and the process tool: General information about a patient is recorded in the cockpit. The process tool displays the current patient dossier and is updated each time the patient visits the hospital. Information that is subject to special secrecy restrictions is entered in the "History" field. This field is only ever visible to the user who submitted the entry, regardless of who is the

[12] I am very grateful to Hubert Kössler and Thomas Wild, the two co-leaders of the chaplaincy team at the Inselspital, for providing me with useful information and for reviewing this chapter.

General situation
- □ Contact offer / request of patient / of relative
- □ Exceptionally stressful situation (emergency care)
- □ Situation of dying

Actions
- □ Procedural support
- □ Existential and/or spiritual support
- □ Religious service
- □ Emergency psychological intervention (emergency care)
- □ Ethical decision support
- □ Triage / networking
- □ Inclusion of relatives

Arrangements
- □ Next conversation: (weekday, date, possibly name)
- □ Check back again (e.g. next week)
- □ No further meetings agreed
- □ Patient / relatives to get in contact if necessary

Fig. 7 i-pdos – Checkboxes Pastoral Care

current dossier owner. If the patient is later transferred to another pastor, the field remains hidden to this chaplain.

The objectives of Seel:is are:

- Collecting basic information: the most important information about pastoral intervention and support is collected and serves as a basis for the continuation of the support and for the personal quality assurance of the work.
- Information transfer: for the further support of a patient by another chaplain (in case of absence) and for the reconstruction of a critical incident.
- Statistical recording: the recorded data is used to record team performance and to evaluate performance according to a wide range of criteria (e.g., presence at certain clinics, involvement with certain groups of people). They thus serve as a justification of the use of funds, a basis for quality development and as material for research projects.

The Seel:is documentation tool will now be presented in more detail using an anonymous case study. Here is the information displayed in the cockpit and process tool (Figs. 8 and 9):

Significantly, following the development of the electronic patient file, the Seel:is will be integrated into the unified documentation system of the "Inselspital."

Fig. 8 Seel:is, Cockpit. (Reproduced with permission from University Hospital Inselspital Bern. Copyright © 2020 University Hospital Inselspital Bern. All rights reserved)

5.3 *Lausanne University Hospital*[13]

The "Centre Universitaire du Vaud (CHUV)" employs 11,364 people and has beds for 1568 patients. The pastoral care team comprises 18 Reformed and Catholic chaplains.

---

[13] I am very grateful to François Rouiller, team leader, and Annette Meyer, member of the chaplaincy team at the CHUV, for providing me with useful information and for reviewing this chapter.

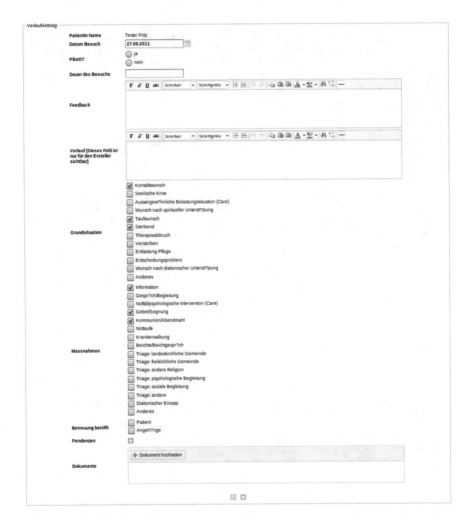

Fig. 9 Seel:is, progress entry. (Reproduced with permission from University Hospital Inselspital Bern. Copyright © 2020 University Hospital Inselspital Bern. All rights reserved)

The electronic documentation of pastoral care was introduced in all clinics 3 years ago (2016). The basic tool is the assessment instrument STIV (French acronym for "Sense, Transcendence, Identity, and Values") and the associated SDAT (French acronym for "Spiritual Distress Assessment Tool"). The two tools are briefly presented below.

5.3.1 STIV and SDAT: The Conceptual Foundations of Documentation

The STIV was developed in the field of geriatrics by the chaplain Etienne Rochat together with the physician Stéfanie Monod-Zorzi in order to support elderly, vulnerable people during their clinical stay in such way that their spiritual dimension would also be taken into account. The STIV is therefore not a general spiritual assessment tool but has its place in hospital life. In this context, the preservation of autonomy and the experience of the meaningfulness of one's own life are recognized as central spiritual needs.

Basic for the STIV is the bio-psycho-socio-spiritual health model, on the basis of which an assessment model was developed which understands the spiritual dimension as an integrative dimension of health: "The patient's spirituality is not only taken into account and evaluated in the same way as the other traditional biological, psychological and social dimensions, but also integrated into the construction of the patient care project" (Monod-Zorzi 2011, 47).

The objectives pursued are to (Monod-Zorzi 2011, 30):

- Promote the autonomy of the person
- Mobilize spiritual resources to better integrate the disease or crisis
- Better understand patient expectations and treatment choices

The STIV model of spiritual needs provides a structured and systematic approach to the appraisal of patients' spirituality. This approach, based on the assessment of patients' spiritual needs, is consistent with the care approaches of medicine and nursing and is therefore understandable to caregivers (Monod-Zorzi 2011, 65). It is assumed that the chaplain is a full member of the treatment team. The interdisciplinary conference stage is therefore crucial, because it represents the real moment of integration of this spiritual dimension with the other traditional bio-psycho and social dimensions (Monod-Zorzi 2011, 59).

Rochat and Monod characterize the spiritual dimension of sick, hospitalized people in the following categories (Monod-Zorzi 2011, 51f):

- Meaning. This concerns the need for a global balance in life, the need to live with a disease or to integrate limitation and disability into life.
- Transcendence. This is about the need for connectedness, which can be expressed in different forms – religious or non-religious.
- Values. This is about a person's value orientation, which comes into play in his or her life decisions. In the hospital, it is a question of these values being recognized and the possibility of maintaining control being maintained (e.g., through involvement in the decisions of the treatment team).
- The psychosocial aspects of identity. This is about the need to maintain one's (unique) identity.

On the basis of the STIV, Monod and Rochat developed a tool for measuring spiritual distress called SDAT for Spiritual Assessment Tool. This tool helps to identify unmet spiritual needs and determine their severity. Spiritual stress is determined with reference to the four categories of the STIV and recorded using a scale from 0 (spiritual needs are not fulfilled at all) to 3 (spiritual needs are fulfilled).

5.3.2   The Documentation Tool

Documentation is provided on two pages: one can only be viewed by the members of the pastoral team, while the other is also accessible to the interprofessional team.

The pastoral page collects general information about the patient (dossier aumônerie), records any interventions, and contains the patient's spiritual anamnesis, which is created using the STIV and is the basis for the spiritual evaluation:

Evaluation STIV:

– Meaning (observations – hypotheses)
– Transcendence (observations – hypotheses)
– Values V1[14] (observations – assumptions)
– Values V2 (observations – assumptions)
– Psycho-social aspects – Identity (observations – hypotheses)

The spiritual evaluation is accessible to the interprofessional team. The evaluation always refers to the current situation and is not a static value. It is determined by assessing which spiritual needs are currently in the foreground for the patient through the evaluation of the four categories. In addition to the spiritual evaluation, the interprofessional page records any recommendations for the treatment team and provides space for further observations and information.

Chaplaincy Point:

– Results of spiritual evaluation (STIV/SDAT)
– Recommendations
– Notes and observations

Efforts are already underway to develop the documentation tool further: The pastoral team is, on the basis of the STIV, currently developing another tool to be complementary to SDAT: the STIV-RePer, which will assess patients' worries (French: PERturbations) and the spiritual resources available to them. The pastoral care team is also currently revising the documentation tool itself, with the aim of (a) reducing time expenditure, (b) giving more weight to spiritual resources, and (c) reducing the room for interpretation.

---

[14]Valeurs 1 are the values that are important for the patient in general. Valeurs 2 are those values that are particularly relevant for the current hospital stay.

5.4 Goals and Current Practice: Charting Systems in Switzerland

In an essay with the programmatic title "Leaving Footprints" (Ruff 1996, 390f), Rob A. Ruff formulated for the first time the central objectives of pastoral documentation:

- Chaplains are integral members of the interdisciplinary care team.
- Documentation will help create more visibility for the chaplain as a professional.
- Documentation of spiritual interventions is proof of compliance with regulatory requirements by the Joint Commission to provide for the spiritual needs of all patients.

Since pastoral care can leave its footprints in the hospital's documentation in all three hospitals presented above, it seems that the objective of integrating chaplains into to interdisciplinary team is being pursued and has been at least partially achieved.

Comparison of the three documentation systems shows that the development of interprofessional cooperation has reached different stages in different hospitals.

The development in the *Triemlispital in Zurich* shows how the hospital has already made it far from self-evident that it is possible for pastoral care to have access to patient documentation. This is probably because pastoral care is not regarded as an integral member of the interprofessional team. Only when the pastoral care team was able to show how the insufficient exchange of information in patient care could actually lead to disabilities and difficulties the hospital management gave permission to the pastoral care team to read and to feed at least some of the information of their pastoral support. Pastoral care is thus enabled to leave footprints by documenting their encounters and recording whether faith, religion, or spirituality is a relevant resource for the patient. However, there is no detailed information about the concrete needs of the patient or the specific interventions of the pastoral care team.

The pastoral care team at the *Inselspital in Bern* has long been an undisputed interprofessional partner in most clinics and thus an integral part of many treatment teams. Consequently, the pastoral counseling department has access to electronic patient documentation in all clinics and can also make entries. It provides information about the patient's situation, about what led to the consultation (contact offer or request, crisis, death situation) and about any pastoral care activities (procedural support, crisis support, ethical support, etc.). In addition, the documentation informs the treatment team about future pastoral care support.

The footprints of pastoral care work are visible in the system and give an approximate idea of the nature of the intervention through the specification of pastoral activities. A special strength of the pastoral documentation of the Inselspital – also when compared to hospitals worldwide – lies in the fact that it generates very precise statistical data and can provide detailed information on a wide variety of interrelationships.

Interprofessional collaboration seems to be most advanced in *some CHUV clinics in Lausanne*. Indeed, the assessment instrument STIV, which is the basis for the clinics' documentation, was developed in interprofessional collaboration. This development took place against the background of long-standing efforts by pastoral care to understand and represent the spiritual dimension as part of health care. These efforts have been supported by cantonal legislation, which makes provisions for the spiritual dimension. In the pastoral care documentation, which the entire treatment team can view, the result of the spiritual evaluation is recorded, and recommendations for the treatment team are formulated.

For this form of documentation, the image of footprints is no longer appropriate. Here, pastoral care does not merely leave traces but intends to participate fully in the healthcare of patients with its own expertise. Its contribution is integrated into the work of the care team, and its assessments are considered by the other health professionals and influence their actions.

Although the documentation systems in the Inselspital and, perhaps most clearly, in the CHUV suggest that pastoral care has now been integrated into interprofessional care, some critical remarks are necessary. The existence of an interprofessional documentation tool does not necessarily mean that interprofessional work is actually carried out. "Paper is patient," they say, and electronic dossiers can just be as patient. Firstly, there is the question of who actually reads the documented pastoral contributions. Secondly, the question arises as to the consequences of interprofessional cooperation and, ultimately, of care support. Are there tools for quality assurance? How does using a documentation tool affect the work of health professionals? Are they properly trained in such a way that they can include the spiritual dimension in their actions? These and other questions will have to be addressed in further research projects if we are to understand the true relevance of pastoral documentation.

6   Summary and Future Perspectives

The survey and the discussions with Swiss chaplains about documentation reveal three things:

- First, there is a great interest in the topic among Swiss chaplains.
- Second, there is an open or positive attitude toward documentation, with many chaplains already documenting, be it for the purposes of interprofessional cooperation, or for the purposes of cooperation within palliative care, and the majority

of chaplains are open to documentation as part of interprofessional collaboration.

– Third, there are worries about how the pastoral secret is affected, whether documentation changes the profile of pastoral care in an unacceptable way (e.g., to medicalize it), and whether it is appropriate to communicate pastoral information in written form, given the possibility of security breaches.

Many *of the factors that have promoted the development of documentation* arose first in the field of palliative care: the fact that interprofessional cooperation, in which pastoral care is integrated, is usually standard in (specialized) palliative care, the requirements of palliative complex treatment as part of financing, and the documentation standards developed by the Pastoral Care Section of palliative.ch. This results in requirements such as joint documentation.

Other factors that promote documentation have to do with the general task and cooperation of pastoral care with other professions in the hospital. Where pastoral care takes over the function of a hospital's internal care team, documentation is important for transfer of cases as well as for the purpose of assessing the quality of the service for the institution: Plausibility check of the service vis-à-vis the institution, the understanding of pastoral care as part of the interprofessional team and thus the standard of interprofessional cooperation as it is increasingly applied in larger hospitals in Switzerland and the need for statistical data to demonstrate the need for pastoral care on the part of the funding institution.

Some factors that hinder the development of pastoral documentation or question its necessity come from within the field of pastoral care and from the church. There is no consensus within the field of pastoral care because there are hardly any developed theories of pastoral documentation, with the consequence that documenting plays no part in pastoral education and training. Meanwhile the position of the churches with regard to documentation varies from canton to canton and from confession to confession.

Other hindering factors come from outside: the legal situation is unclear or controversial, and the political situation varies depending on the canton.

So, looking *to the future,* what should be done? First of all, and most fundamentally, it must be clarified how interprofessional cooperation can or should be structured and whether, or in what form, the documentation is a helpful contribution to this. Beyond that, clarification of the ethical and legal requirements is needed in order to define the possible scope of documentation and provide a clear ethical and legal framework for dealing with the secrecy of pastoral care.

Regardless of how one approaches pastoral documentation, the questions it raises lead to highly exciting and relevant discussions about the future shape of pastoral care in the institutional environment and beyond.

Commentary

Wim Smeets (✉)
Radboud University Medical Center, Nijmegen, The Netherlands
e-mail: Wim.Smeets@radboudumc.nl

There are similarities and differences between the way charting has evolved in Switzerland – as reflected in Pascal Mösli's fascinating contribution – and its development in other countries. The challenges for charting in Switzerland are distinctive in some respects, but have much in common with those faced elsewhere.

1 Similarities

First I will mention the similarities. I see five of them. First of all, charting is developing rapidly in Switzerland, just like in other countries. Until recently only a few hobbyists were interested in an well-structured, shared written documentation of patient contact. Today the entire professional group is dealing with the theme. Second, this recent development is not due to a spontaneous awakening of professional awareness, but is driven by external factors. Just like in the Netherlands, for example, the emergence of integrated palliative care is one of the most important reasons. Palliative teams expect pastors to carefully and richly document their findings and actions, just like other professions do; otherwise the spiritual dimension cannot be properly integrated into healthcare. In our country, the Netherlands, this has even led to the development of a new professional position, the spiritual care consultant: he or she can fully participate in the team – and therefore also report – without being hindered by discussions about professional secrecy. The work of the Taskforce Spiritual Care of *Palliative.ch* and the influence of the "Pastoral Care as Specialized Spiritual Care in Palliative Care" guidelines can be compared with a similar Task Force and the revised guideline "Meaning and Spirituality in Palliative Care" in the Netherlands. It is expected that the application of such guidelines will stimulate the further development of charting. Third, the professional group of (German-speaking) pastors is very divided about whether or not to participate in the interdisciplinary registration of patient contacts. The same division can be seen in other countries; the debate is perhaps equally emotional everywhere. It slows the development of charting and makes it difficult to speak with one voice to companies that develop electronic documentation systems. In this way a mix of systems and documentation methods remains. Fourth, training is an important incentive to discuss data about patient contacts with others. Clinical Pastoral Education has also played a key role in other countries for many years, stimulated by the legacies of Boisen, Gerkin, and other inspirational figures who advocated for learning through living human documents. Fifth, there is the central role of the professional association. It is only since these associations have come to realize the importance of charting that we have seen the development from separate, in themselves valuable, initiatives toward a joint approach.

2 Differences

Then there are the differences. I will mention three of them. In Switzerland, the arrival of a national electronic patient file (the law of April 2017) might be an important stimulator for pastoral documentation, or should we say that the law leaves pastors no choice? In the Netherlands, the parliament has blocked similar initiatives for a national electronic patient file with an appeal to the privacy of the citizen. This aversion to government interference in private life is perhaps something typically Dutch – due to negative experiences during the Second World War – a sentiment that may be difficult to understand for outsiders. A second difference is the influence of emergency care on the development of documentation systems. It would be interesting to hear in more detail how emergency care in Switzerland promoted the professionalization of pastoral care; in the Netherlands we are still at the beginning (Pater, R., W. Smeets, and A. de Vries. 2019. A beacon in a storm: Competencies of healthcare chaplains at the Accident & Emergency Department. *Journal of Healthcare Chaplaincy* [in press]). The biggest difference, however, lies in conducting a national survey among pastors. Although opinions have also been canvassed in other countries, the recent national survey in Switzerland offers a very complete picture of pastors' opinions about charting. The significance of these empirical data, also for the international discussion, cannot be sufficiently stressed. This research is therefore of significant merit! It makes clear, among other things, what a difference it makes whether pastors are employed by institutions or are employed by the church (with variation by canton and by denomination). The largest group of pastors who chart do so on behalf of their institution and with a view to interprofessional cooperation.

3 Challenges

Partly based on the three cases presented by Mösli, five challenges can be formulated, which the Swiss situation shares with those in other countries. The diversity in design of the three examples firstly points to the need for greater uniformity in the reports. The practice of charting often develops pragmatically through the initiative of a few pastors. This approach has the advantage that these "footprints" are closely aligned with what is required for action. The disadvantage is that models depend on personal insights and beliefs and therefore they have a limited scope, as Mösli states: sometimes rather process-oriented, sometimes rather result-oriented, but rarely linked to quality standards, in order that the content of the work would also benefit from reporting. Second, it is desirable that the developed models have a scientific basis. The Lausanne model is a unique example in that regard, because it is based on the Spiritual Needs Model of Stéfanie Monod. This model is related to the Spiritual Needs Model of Fitchett, which we have taken as a starting point in Nijmegen and elsewhere in the Netherlands. With this scientific basis, the randomness of some lists can be avoided. It offers other professionals insight into the scope of pastors' work. Pastoral care is integrated into a holistic view of health. Ruff had arrived at that insight more than 20 years ago, Mösli rightly observes. Third, reporting in this way can be an important building block of research on the profession of pastors and spiritual caregivers. At the moment we are still very dependent on general sociological studies on the religious or spiritual status of the population of a country. With scientifically based reporting systems we can gain a deeper insight into the religious and spiritual attitudes of specific population groups, into the interventions that pastors undertake and the effects thereof. This type of data goes much further than the outcome studies that are currently being set up. Fourth, a full report also provides starting points for aftercare outside the institutions. Aftercare is not possible without a proper report. That is why the Netherlands is now also considering reporting activities by spiritual caregivers and pastors in the home context. This challenge was recognized many years ago by the international Rehabilitation Movement in psychiatry. This makes the documentation by pastors of their actions from the perspective of empowerment indispensable. At the end, Mösli lists what is needed for the further development of reporting. All the elements mentioned are important. What I miss – and this is the fifth and final challenge – is the further development of reporting by and through education and training. Collaboration by pastors among themselves and between pastors and other disciplines would benefit if best practice and critical experiences were shared, or if there were some form of joint supervision. This requires a thorough substantive report on the practical case studies. Working together this way would also encourage each participant in the training to deepen their own reporting. Guidelines only work when training increases knowledge, when skills are practiced and desired attitudes are developed.

The international sharing of reporting insights and practices will improve the quality of reporting systems. The examples presented here, especially the detailed descriptions from

Bern and Lausanne, can provide inspiration for working groups of professional associations to lay alongside their own models. In that sense, the initiative of the Swiss to organize an international meeting of experts and to publish their findings is to be commended. It is now the responsibility of other countries to take up the initiative!

References

Famos, Rita, Matthias Felder, Felix Frey, Matthias Hügli, and Thomas Wild. 2016. *Dem Anvertrauten Sorge tragen – Das Berufsgeheimnis in der Seelsorge*. Bern: SEK/FEPS.

Leitlinien. Seelsorge als spezialisierte Spiritual Care in Palliative Care. 2019. www.palliative.ch/de/fachbereich/fachgruppen/fachgruppe-seelsorge.

Lichter, David. 2015. How do we measure quality in charting? National Association of Catholic Chaplains. *Vision* 2015 (25/4): 3–5.

Monod-Zorzi, Stéfanie. 2011. *Soins aux personnes âgées. Intégrer la spiritualité?* Bruxelles: Lumen Vitae.

Ruff, Robert A. 1996. Leaving Footprints: The practice and benefits of hospital chaplains documenting pastoral care activities in patient's medical records. *Journal of Pastoral Care* 50: 390f.

Spiritual Care in Palliative Care. Leitlinien zur interprofessionellen Praxis. 2018. www.palliative.ch/de/fachbereich/task-forces/spiritual-care.

Charting Spiritual Care: Psychiatric and Psychotherapeutic Aspects

Eckhard Frick

A sick person's spirituality demands a high level of protection, discretion, and confidentiality on the part of the caregiver. At the same time, the caregiver must provide an open and safe setting in order for a patient to feel comfortable addressing his or her spiritual needs or distress. This *interpersonal conflict* between disclosure and non-disclosure depends on professional roles, the spiritual care provider's integration into the healthcare team, and systemic pressure emanating from juridical, ethical, or institutional norms, which usually regulate the patient–caregiver relationship.

In addition, the handling of spiritual confidentiality may provoke an *intrapersonal conflict* in patients, family members, and professional caregivers. Proactively addressing religious and spiritual (r/s) issues has a strong interventional impact on patients. This effect is generally more important than detailed contents gained from spiritual screenings and assessments (Paal et al. 2017):

> Oftentimes they need permission to talk about those kinds of issues. Without some signal from the physicians, patients may feel that these topics are not appropriate or welcome (Puchalski and Romer 2000, 130).

When asked about r/s needs or problems, patients may feel ashamed, bothered, surprised, annoyed, or, conversely, satisfied, supported, and acknowledged in their coping efforts. Consequently, documentation should first and foremost reflect the patient's reaction towards the clinician's r/s intervention and to what extent the patient wishes this interaction to be shared within the healthcare team. There is growing evidence that patients want their caregivers to take into account the spiritual dimension of healthcare (Best et al. 2015). Health professionals must, however, respect individual and general boundaries (non-coercive, non-proselytizing, non-neglecting approach). In psychiatry and psychotherapy, patients' spirituality is less

E. Frick (✉)
Technical University of Munich, Munich, Germany
e-mail: eckhard.frick@tum.de

S. Peng-Keller, D. Neuhold (eds.), *Charting Spiritual Care*,
https://doi.org/10.1007/978-3-030-47070-8_10

pathologized than in former times. In fact, it is increasingly accepted as a universal dimension to human experience, transcending individual religions. In mental health and other medical fields, r/s may be part of the problem, part of the solution, or both (Pargament et al. 2003, 1346). Consequently, spiritual charting should not only differentiate pathological/negative and resilient/positive coping but should also encompass the patient's r/s healthcare preferences and goals, as well as the role he or she attributes to the healthcare professional. A hermeneutical approach is required both when communicating with the patient and when documenting the case for the healthcare team, i.e. when "translating" the patient's spirituality and sharing it with different team members. Team members must, however, always keep in mind their own r/s and their professional affiliations, as well as their experiences and competencies in this field.

1 Some Definitions

Guarding one's *boundaries* – protecting bodily, emotional, and narrative secrets – is not only a human right but a basic anthropological asset. According to Plessner, a physical, inanimate thing (e.g. the objectified anatomical body: "Körper" in German) is distinguished from other things by a border (demarcation or limit) (Plessner 1928/1975, 126). The living body ("Leib" in German), however, is "boundary-realising" ("grenzrealisierend") and has a specific, active relation to this boundary. The living body actively negotiates the "radical conflict between the necessity for closure as a physical body and the necessity for openness as an organism" (218).

From birth, an infant, in its nakedness, seems completely exposed to the caregiver's gaze, attitude, and manipulation. Napkins, clothes, and the caregiver's respect provide boundaries which the infant needs in order to thrive. Slowly, the child learns to invest (psychoanalytically speaking: to cathect) in his or her skin, in the physical and emotional boundaries of his or her own living body. During the second year, the child will learn to feel shame when those boundaries are violated, perhaps only by the other's gaze, be it in a real situation or in his or her imagination. Philosophically speaking, the human being is characterized by its eccentric positionality: it is able to relate to its own centre and "step outside his own physical boundaries by means of reflection. This ability allows man to gain distance from his subjective body (Leib) and perceive it as a separate object (Körper)" (Dobeson 2018, 42). Perceiving or imagining the other's gaze, or reflecting upon oneself entails shame, even in the empathetic embarrassment of "Fremdschämen" (feeling ashamed for somebody else).

Confidentiality protects boundaries and also shame, the leading affect (Leitaffekt) expressing eccentric positionality (Merten 2005; Frick 2015). Confidentiality is a shared attitude between patient and caregiver. We use the term *secret* (i) for everything the patient wants to keep from the knowledge of third parties ("personal secret") and (ii) for the caregiver's confidentiality ("professional secret"). Even matters in no way private may be secret for a person: the term secrecy is different from *privacy*. The professional secret may be individual or collective (*shared secret*). "This difference lies in the means required for maintaining shared secrets. Benjamin Franklin's remark 'Three may keep a secret if two of them are dead' points up the problem" (Bok 1989, 108). Consequently, multiplying the number of confidants may be necessary for facilitating shared teamwork. Nevertheless this sharing requires strict conditions (Liégeois 2010).

Managing privacy and confidentiality means "navigating between the need for autonomy and the need for connectedness with others" (Petronio et al. 2012), i.e. the interpersonal and intrapersonal conflict between disclosure and non-disclosure.

> [...] physicians have potentially two privacy boundaries they regulate with patients. They have their own personal privacy boundaries and judgments about situations where personal disclosures are made to patients. Physicians also serve as guardians or co-owners of their patients' private medical information and are included within the patient's privacy boundary surrounding that information. As co-owners, physicians have a complicated role in that they have to make decisions about issues such as the best treatment plan or a prognosis on the basis of information they gather from tests, and they must deliver that information to the patient. In doing this, physicians often have to judge when to share information with the patient about his/her case, how much to share at any given stage of treatment, what to share, and who else to tell about the patients' confidential medical information (Petronio et al. 2012, 41).

These responsibilities for boundaries apply not only to physicians but also to psychologists and other health professionals. There are three main Communication Privacy Management principles: (1) privacy ownership, (2) privacy control, (3) privacy turbulence.

> Thus, needing to trust a physician to gain health care can motivate a patient to reveal information. For the patient, granting access likely includes judging risks and benefits of allowing complete or partial access. Nevertheless, when the patient discloses information, the physician becomes an authorized co-owner or guardian and, with that, comes an implied expectation that the physician will "care" for the information in the way the patient expects. If these expectations about responsible treatment of the information are violated, *privacy turbulence* [principle #3, E.F.] results in the physician-patient relationship (Petronio et al. 2012, 42).

While confidentiality concerns the helping relationship, *testimonial privilege* refers to investigative authorities. It "protects information possessed by spouses or members of the clergy or lawyers against coerced revelations in court" (Bok 1989, 119).

Not only from a theological but also from an anthropological perspective, it is important to differentiate *secret* and *mystery*. In spoken English, we may use both words as synonyms: we unlock the mystery/secret of someone or something. However, the Cambridge Dictionary defines secret as "a piece of information that is only known by one person or a few people and should not be told to others" and mystery as "something strange or not known that has not yet been explained or understood." Obviously, a secret is concealed by a person and disappears after disclosure. Conversely, the personal origin and the conditions of revealing a mystery remain opaque. Theologically speaking, revelation does not "unlock" but confirms and respects the hidden (absconditus) God. Analogically, knowing the other person in a helping or loving relationship does not eliminate his or her mystery, but deepens it: "As a being exposed to the world, man is concealed to himself – homo abscondi-tus. This concept, originally attributed to the impenetrable essence of God, corre-sponds to the nature of man" (Plessner 1969, 508).

2 Systemic Barriers and Facilitators

Institutional and juridical norms provide a framework for dealing with confidential-ity, preserving patients' rights and balancing caregivers' rights and obligations in this field. In their daily patient work, healthcare professionals continue to be chal-lenged by data protection, data privacy, and the duty of secrecy/obligation to pre-serve confidentiality. In psychiatry and psychotherapy, data are often more sensitive than in other medical fields. A law student, e.g., may be afraid of being disadvan-taged in his or her legal career if a psychotherapeutic consultation is documented.

As far as spiritual care is concerned, charting depends on the spiritual care model observed in a given institutional context, e.g.:

(i) "Spiritual care is not the hospital's business" (Pujol et al. 2016) (French laicism): in public hospitals, spiritual care is delegated to chaplains who work independently of the caring team.

(ii) The board-certified chaplain has the right and the obligation to contribute to electronic health records or other documentation systems (Johnson et al. 2016).

(iii) According to the WHO's definition, palliative care consists of "prevention and relief of suffering by means of early identification and impeccable assessment and treatment of pain and other problems, physical, psychosocial and spiritual" (http://www.who.int/cancer/palliative/definition/en/). This systemic integra-tion of spiritual care is groundbreaking for other medical fields.

(iv) In psychiatry and psychotherapy, the paradigm of pathologizing religion and spirituality is more and more being replaced by efforts to bridge the "'religios-ity gap' between patients, who are more likely to be religious, and psychia-trists, who are more likely to be atheist or agnostic" (Cook and Sims 2018).

In scenario (i), there is neither a need nor an obligation (on the contrary: there is legal prohibition) to take into account spiritual care in the clinical context, at least officially. However, as in other European contexts, spiritual care is an increasingly important subject for "secularised" medicine (Cook et al. 2011) in France (Frick 2006) and in the highly secularized Swiss Canton of Geneva (Huguelet 2017). The integration of a board-certified chaplain (ii) is somewhat the opposite of French secularism. This integration may impose a certain adaptation or "medicalisation" on the chaplain, perhaps putting at risk the independence he or she has in (i). Model (iii) is unique and exemplary, at the same time, for this book's scope: in palliative care, charting encompasses the whole unit of care: patients, families, and caregivers who work to improve the quality of life in an end-of-life context. What does this model mean for psychiatry and psychotherapy (iv)? Not only the World Psychiatric

Organization (Moreira-Almeida et al. 2015) but also national psychiatric organizations, e.g. in Britain (Royal College of Psychiatrists 2013) and Germany (Utsch et al. 2017), strongly recommend fostering spiritual and religious sensitivity among health professionals in psychiatry and psychotherapy.

Liefbroer et al. review two questions addressed in literature (Liefbroer et al. 2019): (i) Who should provide spiritual care – all health professionals as *generalists* (*G*) or chaplains as *specialists* (*S*)?; (ii) What is the role of caregivers' spirituality when providing spiritual care? – the *universalist* (*U*) stance focuses on generic, universal aspects of spiritual care provision and underlines the importance of caring for all patients regardless of the professional caregiver's or receiver's particular spiritual background. Conversely, the *particularist* stance (*P*) highlights the importance of caregivers' own spiritual orientation in spiritual caregiving. Four combinations are possible: *GU*, *GP*, *SU*, and *SP*. Evidently, charting and documentation will be different according to these modes: In *GU*, every team member has a spiritual care responsibility, regardless of personal convictions: a common language for communicating will be necessary. In *GP*, differences between personal convictions in spirituality need to be bridged. In *SU* and *SP*, spiritual care is delegated to a specialist (in general, the chaplain). In the classical confessional mode (*SP*), there will be a tendency for non-disclosure. The *SU*-mode depends on the institutional conditions already mentioned.

3   Code Languages

Professional caregivers are used to specific codes, depending on profession and subprofession, such as internal medicine, surgery, or psychiatry. These code languages are identified empirically, i.e. distinguishing code differences between (sub-)professions is a descriptive, not a normative, judgement. Different professions, e.g. chaplains and nurses, may desire a "universal code" for communicating about patients' spiritual needs and preferences. This desire and an effort to understand the other profession's code may facilitate the sharing of responsibilities. Code differences will, however, not disappear completely. Notably, a complete "totalisation" of languages is not desirable. Furthermore, these codes (often in acronyms, abbreviations, and idioms) are widely used for inside-communication within a given specialty and far less for inter-specialty communication. Palliative care (iii) shares a general outcome-orientation with medical and nursing care. However, as opposed to other medical fields, the outcome here is quality of life, not the elimination of a disease and its roots.

It is true that there is a plethora of studies examining medical and especially psychiatric outcomes in association with religious and spiritual behaviours or activities. Religious involvement is (weakly or moderately) correlated with:

- Purpose and meaning in life
- Higher self-esteem

- Adaptation to bereavement
- Greater social support and less loneliness
- Lower rates of depression and faster recovery from depression
- Lower rates of suicide and fewer positive attitudes towards suicide
- Less anxiety
- Less psychosis and fewer psychotic tendencies
- Lower rates of alcohol and drug use and abuse
- Less delinquency and criminal activity
- Greater marital stability and satisfaction (Koenig et al. 2001, 228)

Health insurances could conclude that a bonus should be given to all insured persons who pray and frequent churches, synagogues, or mosques.

Spirituality, however, is not outcome-oriented. Spiritual codes use words of the health and healing sphere, but in a metaphorical sense. The core concepts of spirituality are the search for the sacred and transcendent. "Hope," e.g., may mean obtaining a negative result in cancer diagnostics. The same word "hope" in the spiritual (transcendent) meaning goes beyond a medically defined outcome, even in unfavourable disease evaluation or in the end-of-life situation. Other examples for different codings are "meaning of life" and "purpose." A psychotherapeutic goal and outcome could be re-establishing a "healthy" meaning and purpose attribution, e.g. in treating depression or narcissistic personality disorder.

Conversely, spirituality is "indifferent" to all those outcomes, as Ignatius of Loyola formulates in his Spiritual Exercises (#23): "[...] so that, on our part, we want not health rather than sickness, riches rather than poverty, honour rather than dishonour, long rather than short life" (Ignatius 1548/1914).

This "indifference" in spiritual coding accepts both meaning in life and meaninglessness, absurdity, grief, and lamentation. These phenomena are not pathology codes. At the same time, spirituality should not be used as a kind of "meaning transplant" towards the suffering person.

In some clinical fields such as palliative care, psychiatry, and psychotherapy, there is a landscape of medical and spiritual codings. We treat diseases and symptoms of body and soul. And even when we "treat the soul" ("mental diseases") we know that with "soul" we name an openness towards transcendence, an openness we cannot "treat" or influence, but only respect.

In certain clinical fields with strong outcome-orientation, e.g. in critical care, chaplains may have a language and a perspective quite different from other caregivers. When they observe and resume their work with terms such as "compassionate presence" (Lee et al. 2017), this may be difficult to understand for other team members and insufficient to illuminate patients' individual stories.

Raffay et al. highlight the diversity of patients' and caregivers' spiritual aspirations (Raffay et al. 2016). They report, nevertheless, the salient principle of co-production ("everyone has a vital contribution to make and brings people who use mental health services, carers, and staff together on equal terms"), which spiritual care encompasses. Furthermore, Heffernan et al.'s review suggests that proactively taking a spiritual history is useful in psychiatry (Heffernan et al. 2014).

4   Possible Interpersonal Conflicts

Given an increasingly diverse religious landscape, health professionals will encounter challenging r/s conversations and new challenges in communication privacy management, as far as the patient's boundaries and their own boundaries are concerned (Canzona et al. 2015). In our own programme for training general practitioners in taking a spiritual history (Straßner et al. 2019) we insist on the capacity to navigate between disclosure and non-disclosure and to respect both the physician's and the patient's boundaries. Furthermore, physicians learn to document what they have discussed when taking the spiritual history. In an interprofessional approach, medical assistants are trained to accompany patients' spiritual evolution in their routine contact.

In 2008, Harold Koenig suggested that good psychiatric practice should include the taking of a spiritual history (Koenig 2008). Koenig's position provoked a vivid controversy in UK psychiatry: This included the critique that spiritual care offers breach proper professional boundaries, that they lack respect for those who reject transcendence, that they open the door to proselytizing, and that they risk causing harm (e.g. to patients with religious delusions). This controversy may be called a conflict between spirituality and secularity in psychiatry (Koenig 2008). In his statement, Cook refers to Taylor's differentiation between a naive ("porous") and a "buffered self" (Taylor 2007):

> In the first condition, the actions of spirits or 'magical' forces is [sic!] sensed as something we experience, as we can the wind, or the elements, or human or animal aggression. The self is 'porous', and one can be to different degrees 'taken over' by such 'magical' agencies. By contrast, for the 'buffered' self, the agency here has to be 'occult', that is, it is something one might accept as a hypothetical cause of events which would themselves be 'naïvely' experienced, such as my falling ill, or suddenly losing a capacity I counted on (Taylor 2010, 415).

For Taylor, the buffered secular self entails "a mutual fragilisation of different positions, and the resulting sense of optionality." The buffered self is cut off from transcendence, restricted to an immanent frame.

Liégeois recommends a model of "conditional shared professional secrecy" (Liégeois 2010). His spiritual care model is the specialist-particular/universal (*S-P/U*). Nevertheless, for practical and ethical reasons, his arguments are helpful for other spiritual care models, too. Not very different from the communication privacy management model (Petronio et al. 2012; Canzona et al. 2015), he addresses the ethical, juridical, and spiritual conflict between confidentiality/non-disclosure and team communication/disclosure and formulates five conditions: (1) the care professionals participate in a clearly defined and identifiable team or network, (2) they perform a common task in care, (3) they are bound by the duty of professional secrecy, (4) they engage in a consultation to ask the patient's informed consent, and (5) they apply the filter of relevance.

The *filter of relevance* is particularly important not only for team meetings and oral communication, but also for charting, for electronic documentation, etc. The spiritual caregiver has to *choose* information of clinical interest. He/she may also check with the patient which information he/she wishes to be disclosed to the caring team.

5   Possible Intrapersonal Conflicts

Coping with conflicts, navigating between and balancing spiritual confidentiality and disclosure may provoke shame in an interpersonal or group setting, e.g. when the team or a reporting team member experiences voyeurism, perhaps in an ambivalence between avoiding and desiring it.

As a matter of course, shame is still more salient when a person such as the patient is vulnerable. In a dual relationship, shame may be tolerated or it may even permit a certain intimacy, e.g. in basic nursing acts or in a respectful medical examination.

As we saw, the human being is "boundary-realising" and may shamefully perceive the fragility of its body and soul. Respecting shame is of immense importance not only in a romantic relationship but also in every helping or therapeutical professional relationship.

When boundaries are threatened or need protection, shame arises: this is particularly true for the health professionals' and patients' boundaries in the sphere of religion and spirituality. According to Charles Taylor, the secular self loses its "porosity" and protects itself by "buffering" and by multiplying spiritual options (Taylor 2007). When the patient's spiritual quest is not only listened to (in a conversation with his or her doctor), but also communicated within the caring team or even documented in a chart, this may provoke "porosity-anxiety" on the part of the patient.

6   Mystery Is More Important than Mastery

Contemporary healthcare must get to grips with numerous problems, which humanity has faced for centuries without the hope of healing or even alleviation. To this end, it is important that our intention is to understand a sick person and to share this understanding in the caring team.

What the paediatrician and psychoanalyst Winnicott calls "incommunicado" (Winnicott 1960/1990, 186) is, anthropologically speaking, the mystery of the human being (Plessner 1969). In this contribution, I have reflected upon the conflicts between non-disclosure and disclosure in spiritual care, and upon a responsible boundary management. The ultimate boundary protects the self's core, the incommunicado.

References

Best, Megan, Phyllis Butow, and Ian Olver. 2015. Do patients want doctors to talk about spirituality? A systematic literature review. *Patient Education and Counseling* 98 (11): 1320–1328. https://doi.org/10.1016/j.pec.2015.04.017.

Bok, Sissela. 1989. *Secrets: On the ethics of concealment and revelation.* New York: Vintage.

Canzona, Mollie Rose, Emily Bylund Peterson, Melinda M. Villagran, and Dean A. Seehusen. 2015. Constructing and communicating privacy boundaries: How family medicine physicians manage patient requests for religious disclosure in the clinical interaction. *Health Communication* 30 (10): 1001–1012. https://doi.org/10.1080/10410236.2014.913222.

Cook, Christopher C.H., Andrew Powell, Andrew Sims, and Sarah Eagger. 2011. Spirituality and secularity: Professional boundaries in psychiatry. *Mental Health, Religion and Culture* 14 (1): 35–42.

Cook, Christopher C.H., and Andrew Sims. 2018. Spiritual aspects of management. In *Textbook of cultural psychiatry*, ed. Dinesh Bhugra and Kamaldeep Bhui, 472–481. Cambridge: Cambridge University Press.

Dobeson, Alexander. 2018. Between openness and closure: Helmuth plessner and the boundaries of social life. *Journal of Classical Sociology* 18 (1): 36–54. https://doi.org/10.1177/1468795x17704786.

Frick, Eckhard. 2006. Peut-on quantifier la spiritualité? Un regard d'outre-Rhin à propos de l'actuelle discussion française sur la place du spirituel en psycho-oncologie. *Revue Francophone de Psycho-Oncologie* 5 (3): 160–164.

———. 2015. Psychosomatische Anthropologie. Ein Lern- und Arbeitsbuch für Unterricht und Studium *(2. Auflage)*. Stuttgart: Kohlhammer.

Heffernan, Suzanne, Sandra Neil, and Stephen Weatherhead. 2014. Religion in inpatient mental health: A narrative review. *Mental Health Review Journal* 19 (4): 221–236. https://doi.org/10.1108/MHRJ-09-2014-0035.

Huguelet, Philippe. 2017. Psychiatry and religion: A perspective on meaning. *Mental Health, Religion & Culture* 20 (6): 567–572. https://doi.org/10.1080/13674676.2017.1377956.

Ignatius. 1548/1914. *The spiritual exercises of St. Ignatius of Loyola (translated from the autograph by Elder Mullan, s.j.).* New York: P.J. Kenedy & Sons.

Johnson, Rebecca, M. Jeanne Wirpsa, Lara Boyken, Matthew Sakumoto, George Handzo, Abel Kho, and Linda Emanuel. 2016. Communicating chaplains' care: Narrative documentation in a neuroscience-spine intensive care unit. *Journal of Health Care Chaplaincy* 22 (4): 133–150. https://doi.org/10.1080/08854726.2016.1154717.

Koenig, Harold G. 2008. Religion and mental health: What should psychiatrists do? *Psychiatric Bulletin* 32: 201–203.

Koenig, Harold G., Michael E. McCullough, and David B. Larson. 2001. *Handbook of religion and health.* Oxford: Oxford University Press.

Lee, Brittany M., Farr A. Curlin, and Philip J. Choi. 2017. Documenting presence: A descriptive study of chaplain notes in the intensive care unit. *Palliative & Supportive Care* 15 (2): 190–196. https://doi.org/10.1017/S1478951516000407.

Liefbroer, Anke I., R. Ruard Ganzevoort, and Erik Olsman. 2019. Addressing the spiritual domain in a plural society: What is the best mode of integrating spiritual care into healthcare? *Mental Health, Religion & Culture*: 1–17. https://doi.org/10.1080/13674676.2019.1590806.

Liégeois, Axel. 2010. Le conseiller spirituel et le partage d'informations en soins de santé. Un plaidoyer pour un secret professionnel partagé. *Counselling and Spirituality / Counseling et spiritualité* 29 (2): 85–97.

Merten, Jörg. 2005. Facial microbehavior and the emotional quality of the therapeutic relationship. *Psychotherapy Research* 15 (3): 325–333. https://doi.org/10.1080/10503300500091272.

Moreira-Almeida, Alexander, Avdesh Sharma, Bernard Janse van Rensburg, Peter J. Verhagen, and Christopher C.H. Cook. 2015. *WPA Position statement on spirituality and religion in psychiatry.*

http://www.wpanet.org/uploads/Position_Statement/WPA%20position%20Spirituality%20 statement%20final%20version_rev2%20on%20Spirituality.pdf, access 8.5.2016.

Paal, Piret, Eckhard Frick, Traugott Roser, and Guy Jobin. 2017. Expert discussion on taking a spiritual history. *Journal of Palliative Care* 32 (1): 19–25. https://doi.org/ 10.1177/0825859717710888.

Pargament, Kenneth I., Brian J. Zinnbauer, Allie B. Scott, Eric M. Butter, Jill Zerowin, and Patricia Stanik. 2003. Red flags and religious coping: Identifying some religious warning signs among people in crisis. *Journal of Clinical Psychology* 59 (12): 1335–1348.

Petronio, Sandra, Mark J. Dicorcia, and Ashley Duggan. 2012. Navigating ethics of physician-patient confidentiality: A communication privacy management analysis. *The Permanente Journal* 16 (4): 41–45. https://doi.org/10.7812/tpp/12-042.

Plessner, Helmuth. 1928/1975. *Die Stufen des Organischen und der Mensch. Einleitung in die philosophische Anthropologie*. Berlin: De Gruyter.

———. 1969. De homine abscondito. *Social Research* 36 (4): 497–509.

Puchalski, Christina M., and Anna L. Romer. 2000. Taking a spiritual history allows clinicians to understand patients more fully. *Journal of Palliative Medicine* 3 (1): 129–137.

Pujol, Nicolas, Guy Jobin, and Sadek Beloucif. 2016. 'Spiritual care is not the hospital's business': A qualitative study on the perspectives of patients about the integration of spirituality in healthcare settings. *Journal of Medical Ethics* 62 (11): 733–737. https://doi.org/10.1136/ medethics-2016-103565.

Raffay, Julian, Emily Wood, and Andrew Todd. 2016. Service user views of spiritual and pastoral care (chaplaincy) in NHS mental health services: A co-produced constructivist grounded theory investigation. *BMC Psychiatry* 16: 200–200. https://doi.org/10.1186/s12888-016-0903-9.

Royal College of Psychiatrists. 2013. Recommendations for psychiatrists on spirituality and religion, position statement PS03/2013.. Royal College of Psychiatrists.

Straßner, Cornelia, Eckhard Frick, Gabriele Stotz-Ingenlath, Nicola Buhlinger-Göpfarth, Joachim Szecsenyi, Johannes Krisam, Friederike Schalhorn, Jan Valentini, Regina Stolz, and Stefanie Joos. 2019. Holistic care program for elderly patients to integrate spiritual needs, social activity, and self-care into disease management in primary care (HoPES3): Study protocol for a cluster-randomized trial. *Trials* 20 (1): 364–364. https://doi.org/10.1186/s13063-019-3435-z.

Taylor, Charles. 2007. *A secular age*. Cambridge, MA: Belknap Press of Harvard University Press.

———. 2010. Challenging issues about the secular age. *Modern Theology* 26 (3): 404–416.

Utsch, Michael, Ulrike Anderssen-Reuster, Eckhard Frick, Werner Gross, Sebastian Murken, Meryam Schouler-Ocak, and Gabriele Stotz-Ingenlath. 2017. Empfehlungen zum Umgang mit Religiosität und Spiritualität in Psychiatrie und Psychotherapie. *Spiritual Care* 6 (1): 141–146. https://doi.org/10.1515/spircare-2016-0220.

Winnicott, Donald Woods. 1960/1990. *The maturational process and the facilitating environment*. London: Karnac Books & The Institute of Psychoanalysis.

Palliative Chaplain Spiritual Assessment Progress Notes

Paul Galchutt and Judy Connolly

1 Background

Palliative care is a field of medicine focused on caring for people with serious illness. Unique to palliative care is the interprofessional team delivery model. In the United States, the team typically consists of physicians, non-physician prescribing and treatment providers (e.g., advanced practice nurses), psychotherapists (e.g., clinical social workers), and a chaplain. The holistic nature of palliative care is summarized in eight domains (Ferrell et al. 2018). The fifth domain, relating to existential, religious, and spiritual aspects, is the area in which the chaplain conceptually serves as the specialist. Based on this fifth domain, the *National Consensus Project Clinical Practice Guidelines for Quality Palliative Care* (2013) cites, "Preferred Practice 20: Develop and document a plan based on assessment of religious, spiritual, and existential concerns using a structured instrument and integrate the information obtained from the assessment into the palliative care plan" (p. 64).

Palliative care chaplains contribute to this plan through spiritual assessment. A fairly recent white paper for palliative spirituality (Puchalski et al. 2009) highlights the importance of the spiritual care role for all palliative interprofessionals. It also makes a distinction between the specialist role of chaplains, focused on the patients' existential, spiritual, and religious domain, and the generalist role of non-chaplain palliative team members such as physicians, advanced practice providers, and clinical social workers (Puchalski et al. 2009).

Spiritual assessment is an essential component of healthcare chaplaincy in the United States. The US-based Association of Professional Chaplains lists spiritual assessment as the first standard, among 15, *Standards of Practice for Professional Chaplains.* It states, "The chaplain gathers and evaluates relevant information regarding the care recipient's spiritual, religious, emotional and relational needs and

P. Galchutt (✉) · J. Connolly
M Health Fairview, Minneapolis, MN, USA
e-mail: pgalchu1@fairview.org

© The Author(s) 2020
S. Peng-Keller, D. Neuhold (eds.), *Charting Spiritual Care*,
https://doi.org/10.1007/978-3-030-47070-8_11

resources" (2015, p. 1). This standards document grants flexibility for the form in which a chaplain records this information about a spiritual assessment into the patient electronic medical record (EMR). For some healthcare chaplains, spiritual assessment can mean a diagnosis developed, for example, around the issue of belonging or forgiveness. For other chaplains, spiritual assessment can mean inputting point-and-click logistical data in the EMR concerning where the visit occurred, how long it lasted, or the basic interventions facilitated. There is some evidence (Tartaglia et al. 2016), however, that chaplains prefer to transmit the information through narrative text or what is referred to as the progress note in the EMR. Our contribution to this volume is focused on the spiritual assessment conducted by a palliative care chaplain as it takes the form of a narrative progress note. One purpose of this progress note is to contribute to the palliative care team's shared care plan for the patient.

This plan is collaboratively developed with the chaplain and interprofessional partners on the palliative care team based on a structured spiritual assessment instrument. The palliative spirituality white paper (Puchalski et al. 2009), which was developed as part of a consensus project, stated that chaplains' spiritual assessments are models of "interpretive frameworks [...] based on listening to a patient's story as it unfolds" (p. 893). Over the years, many of these spiritual assessment models have been developed by chaplains, for chaplains (Fitchett 2017) to serve as templates for narrative progress note completion.

Some evaluation, albeit limited and sporadic, of healthcare chaplains' spiritual assessment progress notes exists from non-chaplains. A Duke University team (Choi et al. 2019) studied chaplain progress notes as a component of an intensive care unit setting which featured the views of non-chaplain clinicians. They reported a prevalence of attending physicians (31%) and nurses (23%) reading chaplain progress notes (p. 94). While not overly large sample sizes (attending physicians, n = 29; nurses, n = 139), these data are significant due to the scarcity of such data anywhere within the peer-reviewed literature. Also noteworthy are observations made by another Duke University team (2016) of a medical student (Lee) and two physicians (Curlin and Choi), in their descriptive study of chaplain progress notes also from within an intensive care unit. They wrote of chaplains using "code language" that does not "convey the deeper spiritual connections" (Lee et al. 2016, 194) chaplains have with patients and families. They also note the importance of "standardization" (p. 195) and how it is lacking within chaplain progress notes documented in the EMR.

While chaplains are recognized for their specialized spiritual assessment role within palliative care, a gap remains between what and how a chaplain reports this information through progress notes as well as how it is relevant to the overall care plan. Knowing there is limited evaluation in the literature from non-chaplains regarding chaplain spiritual assessment progress notes, a need exists for a systematic data collection and analysis of the views and perceptions from non-chaplain palliative care team members as the primary readers of these notes. The primary aim then of this study was to ask palliative non-chaplain team members what is most helpful as well as missing from a chaplain's spiritual assessment progress note.

To facilitate this aim to glean the perceptions and perspectives of non-chaplains, seven focus groups were conducted.

The anticipated benefits of these focus groups were quality improvement through knowledge gained or insights offered for what a chaplain communicates to palliative care team members. A palliative chaplain's spiritual assessment progress notes are created through and informed by a narrative approach (Puchalski et al. 2009) while making a deeper spiritual connection (Lee et al. 2016). Each patient's values (Frank 2010) will more likely emerge through knowing each person (Cassell 2004). We conducted a descriptive, exploratory study of seven focus groups with the primary aim to discover what content is most helpful as well as missing from the perspective of non-chaplain palliative care team members. This study identifies both descriptive and summary content themes to address the gap between what a chaplain reports through the progress note and how it is relevant to the most heightened needs of non-chaplain palliative providers.

2   Study Design and Data Collection

Having had the rich experience of being an inpatient palliative care chaplain for 10 years, Paul has firsthand concrete practice with seeking to communicate information in a progress note not knowing if it was being read and, if read, whether it was making a difference. Motivation behind this project was to ask, essentially, what information can best help chaplains make a relevant and meaningful difference together with palliative care team members to reduce suffering and improve quality of life.

We conducted focus groups with non-chaplain palliative care team members to address the research question: What content is most helpful as well as missing from palliative chaplain spiritual assessment progress notes as perceived by non-chaplain palliative care team members?

90-minute focus groups were conducted approximately once every 10 days over a span of roughly 3 months from September to November 2018. Participants were from six palliative care teams based out of six acute care hospital locations within a metropolitan area within the upper Midwest region of the United States. One of the six teams was based out of a children's hospital. Two focus groups were hosted in this pediatric setting. The other five focus groups were hosted with the inpatient palliative care teams caring for adult patients.

Each palliative care team consisted, primarily, of physicians, non-physician prescribing and treatment providers (e.g., advanced practice nurses), psychotherapists (e.g., clinical social workers), and nurses such as care coordinators. Also participating was a child-family life specialist as well as a massage therapist. A discussion guide (Table 1) was used to invite the views and perceptions of participants as the focus group conversations unfolded.

During the data analysis processes, we developed concepts and categories to identify patterns and trends by comparing "one segment of data with another to identify similarities and differences" (Krueger and Casey 2015, 157). As the analysis

Table 1 Key focus group questions

1.	Read Introduction. Participant introductions along with opening question – would you share your name and how long have you been involved in palliative care?
2.	So, now we'd like for you all to share with each other, what do you think about the chaplain's spiritual assessment? (Probes: Do you read it? How often?)
3.	Tell us about what is missing from the chaplain's spiritual assessment progress note.
4.	Think back to over the months or years you've been in palliative care and tell us about a memory that stands out as useful from something you've read in a chaplain's spiritual assessment progress note.
5.	Let's talk about what content is in the spiritual assessment progress note. (a) Take this piece of paper and write down which three things from the spiritual assessment are most helpful to you in your work. (b) OK. Let's go around the table and each of you tell me what you wrote down and give me a one-sentence description of that thing. (List each item on a flip chart. If an item is mentioned more than one time, put a check mark next to it for each additional time it is mentioned.) (c) (Pick the one with the most check marks and say) A number of you said X was a strength. Talk more about that. (Discuss two or three items – as time allows.)
6.	Now, let's talk about what's missing from the spiritual assessment progress note. Let's use the same process. (a) Using the same piece of paper, write down three things missing from the spiritual assessment. (b) OK. Let's go around the table and each of you tell me what you wrote down and give me a one-sentence description of that thing. (List each item on a flip chart. If an item is mentioned more than one time, put a check mark next to it for each additional time it is mentioned.) (c) (Pick the one with the most check marks and say) A number of you said X was missing. Talk about that. What about X is important to the spiritual assessment and your work with the chaplain. (Discuss two or three things as time allows.)
7.	So, we've talked about features within the spiritual assessment progress note. Thinking about all of those, I'd like for each of you to now share what you see as the most important thing we've discussed and tell us why you feel that way.
8.	Conclusion: Oral summary of focus group provided by the assistant moderator

partners, we then met again to compare and contrast the groups framed by these categories for a second stage of coding as well as to strengthen reliability. For the next stage of the constant comparative method of theme development, defined themes with representative quotes from the data were proposed based upon the initial categories. The analysis partners, finally, confirmed theme selection with one another prior to manuscript creation to further enhance reliability of the data.

In total, 42 non-chaplain, palliative care interprofessionals participated in the 7 focus groups occurring during Fall 2018. Focus group sizes ranged from 4 to 8 participants (mean = 6). Table 2 displays the demographic characteristics. Participant ages were between 25 and 62 years of age (mean age 43). Most participants were white (91%) and female (69%). Among the interprofessional participants are physicians (41%), non-physician prescribing and treatment providers (21%), nurses (17%), and psychotherapists (17%). Although many participants are identified as Christian (62%), there was a noteworthy prevalence (16%) with no religious or spiritual affiliation. While this was also an experienced assembly of healthcare pro-

Table 2 Characteristics of non-chaplain interprofessional palliative care team members (N = 42)

Characteristic	Total (n = 42) $f(\%)$
Vocation	
Physician	17 (41)
Advanced practice provider[a]	9 (21)
Registered nurse	7 (17)
Clinical social worker	7 (17)
Child-family life specialist	1 (2)
Massage therapist	1 (2)
Age	
Mean = 43 years old	12 (29)
25–36	18 (42)
37–49	12 (29)
50–62	
Palliative work experience	
<1 year – 9 years	29 (69)
10–19 years	11 (26)
20–28 years	2 (5)
Gender	
Female	29 (69)
Male	12 (29)
Gender non-conforming	1 (2)
Race	
White	38 (91)
African American	2 (5)
Hispanic	1 (2)
Korean American	1 (2)
Religion	
Christian	26 (62)
No faith/spirituality indicated	7 (16)
Jewish	3 (7)
Unitarian Universalist	2 (5)
Spiritual, but not religious	2 (5)
Agnostic	2 (5)

[a] Advanced practice provider (non-physician healthcare provider with scripting authority. Either an advanced practice nurse or a clinical nurse specialist within this sample)

fessionals with a continuum beginning with less than 1 year of experience extending to 28 years, the highest percentage (69%) of participants have been serving in palliative care for 9 years or less.

3 Results

Twelve themes, in total, arose from the seven focus groups in response to the research question concerning what content is most helpful as well as missing from palliative chaplain spiritual assessment progress notes. As shown in Table 3, these

Table 3 Themes for palliative chaplain progress notes – focus group research "…the things that are missing are the things that are helpful"

Descriptive content	Summary content

1. Decision-making
   Religion/spirituality
   Hope
	1. Logistics
	   Why there?
	   Time spent?

2. Suffering Religion/spirituality	2. Synthesis

3. Coping
   Religion/spirituality
	3. Scales
	   Suffering/coping
	   Decision-making

4. Religion/spirituality
   Spectrum
   Description
   Importance
	4. Recommendations to staff
	   Language
	   Religious/spiritual practices

5. Story
   Understanding of illness
   Spiritual story
	5. Needs/goals of care/action plan

6. Family
   Support
   Dynamics	

7. Perception of emotion	

12 themes and subthemes were categorized into 2 overall groupings, Descriptive Content (7 themes) and Summary Content (5 themes). These themes are organized by what was both simultaneously helpful and missing as a research participant expressed "…the things that are missing are the things that are helpful."

3.1 Descriptive Content Themes

Through our analysis, seven themes and supporting subthemes emerged related to the descriptive content non-chaplain palliative care team members believe should be addressed and incorporated into a palliative chaplain's progress note. Of note, while the theme of religion and spirituality is specified as a distinct descriptive theme based on the participant's views, it also frequently appeared as a subtheme often integrated into the other themes discussed. Krueger and Casey (2015) write about analytic factors such as extensiveness, frequency, specificity, and emotion related to focus group data being easier to discern from the text if the focus group moderators are also the research analysts as was true with this research project.

3.1.1 Decision-Making

By serving in both the moderator and analysis roles, it was clear that decision-making was the most discussed theme across all groups (extensiveness) and mentioned among all the themes with the most frequency within each group.

Decision-making also prompted the most stories (specificity) and generated the most passion (emotion), especially within the context of palliative care teams facilitating goals of care discussions concerning end of life when conflicting values between family members emerged or when the primary medical team's recommended course for treatment was not being followed.

Decision-making was also discussed as the area participants felt the most pressure about and about which the stakes were often highest within life-and-death patient decision-making situations. They explained a chaplain's detailed reporting about the subtheme of a patient's religion/spirituality and how or why it conflicts with a medically recommend course as being most helpful. Participants further explained having conversations about these matters with chaplains in team rounds but that it was not in the progress note as often as desired. One participant also named the significance for not only knowing religious or cultural information for actionable decision-making but also for personal understanding of the patient and family: "It's very important to know, you know, how one's religion or culture affects their decision making is paramount in understanding."

The participants were also knowledgeable about chaplains inviting conversation about and receiving information from patients and families about their sense of future amid serious illness. Participants expressed their expectation that a chaplain would be descriptively attentive to this information in a progress note when a patient shared their sense of hope or future, as a second subtheme, related to decision-making. One participant simply and specifically named: "Like what is their hope and what are they hoping for."

3.1.2 Suffering

With the relief of suffering or distress being central to the principles of palliative care, it makes sense that palliative care chaplains are expected, based on these focus groups, to address this theme as essential to palliative care in chaplaincy spiritual assessment. In addition to suffering being addressed from a whole person perspective such as paying attention to "trauma, grief, and loss," it was also clear that chaplains are expected to be specialists concerning religious and/or spiritual suffering and to have that reflected in the progress note.

The wish for chaplaincy notation about a patient's overall sense of suffering emerged, including what this means related to physical symptoms. A participant remarked on wanting information concerning this data in a chaplain's note:

> Not that I go to the chaplain, primarily, for their thoughts on the patient's comfort level, but it is helpful sometimes if they…can identify at least if somebody is really in distress. Like maybe, because of symptoms, maybe a combination….

Palliative team members also discussed recognizing that suffering has various dimensions, along with or beyond the physical pain or other forms of bodily distress. The subtheme of religious or spiritual suffering was discussed as palliative care team members as spiritual generalists are aware of and look for information in the chaplain notes about "some kind of spiritual issue."

3.1.3 Coping

Coping exists together with suffering/distress. Focus group participants often reflected on how it is helpful to read in a progress note about some aspect of how a patient is adjusting or what their "strengths" are, overall, related to their serious illness circumstances and what it means for them in terms of their personal suffering/distress. One palliative clinician specified:

> ...the chaplain's perspective on how the patient copes and how they are coping, and kind of where they're at in this process. It's, that's valuable information.

There was also substantive conversation among the teams regarding the subtheme of a chaplain's assessment of, when pertinent, a patient's religious/spiritual sense of coping. A descriptive sense of what orients a patient or family member to that which is sacred or significant during a disorienting and uncertain time of serious illness was named as helpful. Also discussed was how this potentially positive source of coping can or has changed during illness. This sometimes appeared with non-chaplain team members also wanting to know a patient's view of dying or whether they are "at peace." One participant requested this more transcendent data: "...where is this patient spiritually, meaning, 'Are they at peace with where they are spiritually?' Is their spiritual place helpful or harmful to them right now?"

3.1.4 Religion/Spirituality

This thematic area is central to palliative care chaplaincy. There were no efforts made by interprofessional participants to separately define the constructs of spirituality and religion within the context of these focus groups. There was affirmation of palliative chaplains addressing a more global sense of a person's spirituality with three related, but distinct, subthemes. First, palliative team members want information about where a patient is on a "spiritual spectrum" with religion being included within that construct. One participant suggested:

> I'd like to know where they fall on the spiritual spectrum. Like faith is everything versus there is no God versus I'm a spiritual person, I love nature. I want to know where they are in that realm.

While wanting to know matters such as a specific religion or where a person worships, participants asked for information on how religion or their version of spirituality shapes their "view of the world...into a way to make sense of it all." The second subtheme, description, was illustrated when the focus group conversations devoted energy to wanting a description of a patient's religion/spirituality and practices and how it "influences their decision-making," suffering, and coping, as noted above.

The third subtheme of the importance of a patient's religion or spirituality amid serious illness was named as being useful, especially concerning how it informed interprofessional members about the intensity of a worldview or belief system. One participant expressed the significance of this weighted "need" in a chaplain's note, "to dig into the nitty-gritty on the core issue of how is this person's belief of faith or spirituality or religion affecting their current condition." Reflecting the importance of this subtheme, stories were shared in the focus groups illustrating memorable occurrences for both when a patient's religion or spirituality was helpfully detailed in a chaplain's progress note and when it was not.

3.1.5   Story

It was noted by participants that being able to glean a sense of who a patient is as a person through story is helpful. With the diagnosis of any serious illness, a reconfigured story of a person's life as patient necessarily emerges. An unbidden illness has radically changed and disrupted a life. The telling of stories is often the process a patient reflexively uses to make sense of and reconstruct his/her life with serious illness and, sometimes, dying. A palliative chaplain is expected to be a "story-listener" and to integrate an appreciable interpretation and then translation of the meanings and values from these stories into, first, an understanding of illness and, second, when religion/spirituality is identifiably present, the reporting of how his/her "spiritual story" influences illness apprehension.

When talking about gathering a sense of a patient's understanding of illness through a chaplain's note, one participant stated, "I think they're crucial to painting out the whole picture in patients like this." Another participant described not wanting an overly detailed templated note, wanting instead the essence of a patient's understanding, "It's just too much words. I just like to read a story rather than trying to read all of those same questions over and over again."

Consistent with other themes above, religion/spirituality also informed the subtheme of wanting a patient's "spiritual story." It was best summarized by one participant:

> I like reading stories and so I would be interested in a spiritual story, not just the narrative you glean but what is there. Like how they've framed the spiritual story. An explanation of how a patient interprets their illness through a spiritual lens.

3.1.6   Family

The focus group participants indicated that it is helpful to know who the patient defines as family, whether they are involved or influential, and how supportive they are as a subtheme. Related to having a sense of who supports the patient, one of the interprofessional team members requested:

> So, who's in the room, who's coming to visit this [person] and who is a potential influence? And, what are the influences?

It was also affirmed that this support sometimes comes in the form of a patient's religious or spiritual communities in "very concrete and tangible" form as more of a "kind of existential effect."

Participants also want it identified who is a source of conflict or produces tension in manifesting family complexity and dynamics as another subtheme. One participant noted finding this information, historically, in a chaplain's progress note and finding it relevant:

> I think in the family dynamics, sometimes the spiritual care notes can have some insight into, this is the way this family comes to decisions, or this, which can be super helpful in how to approach that family in the care conference.

As is often the case with family dynamics, especially in matters of intense life-and-death decision-making, serious illness is inescapably fraught with layers of emotion.

3.1.7   Perception of Emotion

Conversation in the groups also steered toward a desire for chaplains to communicate some sense of what was perceived emotionally in the room to make the interprofessional partners aware and provide some context. Whether a seriously ill patient and family are integrating a difficult diagnosis and prognosis in a way that is accepting of the medical characterization of it, or whether there is a choosing of a future version of reality that conflicts with the medical picture being painted, perceptions of that information and emotions connected with it were desired. When inpatient palliative care is consulted with the more challenging and conflicting circumstances regarding decision-making, a chaplain's insights are sought to help describe emotional tensions and create awareness for other team members. One participant narrated how this awareness is helpful to framing his/her approach with the patient:

> What are you thinking when you walk into that room? What do you feel? Because I'll probably feel that same thing when I walk to that door.

3.2   Summary Content Themes

Connected with the seven descriptive content themes above, participants also talked about it being helpful for this content to be readily and easily accessible in a summary form in a chaplain's progress note. They expressed wanting the option to read the substantive information in the more descriptive context found in "paragraphs," as well as needing information that was immediately accessible at a glance that was "succinct" or in "bullet points." One of the participants characterized and summarized this need and challenge for both descriptive and summary content:

> ...one of the more challenging parts of writing a note is to figure out, to take what you've gleaned in narrative from a visit and trying to distill it down to a few points that you think that the person reading the note is going to need to understand. That's why editors write headlines and journalists write paragraphs generally [laughter].

Another participant provided a sense of what a chaplain's summary could include:

> You know, the two-line, and again I'm kind of thinking about spiritual care writ large that, you know, 'came and prayed and offered support,' that doesn't tell me very much. But the 2-3-line summaries that kind of indicates the intersection of spirituality and medical care and decision making what they know, that is helpful.

The need for summary content is shown through the following five themes and subthemes. There are echoes of the seven descriptive content themes, weighted, within the summary content themes.

3.2.1 Logistics

When asking participants about what is helpful or missing, some participants responded with an ever-practical logistical request for the citing of a reason for chaplaincy involvement to be front and center as the subtheme of "why there." A participant requested it as being helpful to know, "…why they are being seen? Like, how did the chaplaincy get involved or spiritual care get involved with it?"

Additionally, participants expressed the desire for a chaplain's notation around time spent for establishing an estimation of the possible intensity or involvement of the chaplain encounter. One participant reflected on what this information could mean for his/her care with a patient:

> I mean I do find it helpful for when one of my colleagues go like document prolonged care time, from here to here. I know wow, they really dug in there. And so, if I had some sense maybe of the time spent there, I would maybe know from my end how much, should I, or do I need to tack onto that or if [chaplain] spent two hours there maybe I don't need to be super diving in or I don't know.

3.2.2 Synthesis

Two reasons appeared from the data about the significance of a synthesis at the top in the summary section of a palliative chaplain's note. The first is that it is practical to have the most important information about a care encounter, "the really important stuff," be listed in the space where a busy partnering clinician may be more likely to read it, especially if the information culled out has been discerned to be significant. This leads to the second observation that if the information was assessed to be most important, it likely went through a process of refining discernment and consideration reflected in the quote below:

> When I read a synthesized note, I'm fairly certain that the person has really thought it through and really can put words to what this thing means. And so, there's an aspect of sureness of the information and when I see that, I'm really ok, this is really a very solid bedrock.

3.2.3 Scale(s)

Participants also wanted scales or some kind of a ranking or score concerning both a patient's suffering/coping continuum and decision-making to be immediately accessible in a summary section. It was discussed that one of the functions or a purpose for having a scale is to triage patients with the highest or most intense need. It is important to note that while there was a desire for a quick or easy, readable, and interpretable scale, it was also acknowledged that there may not be a quick or easy solution to the complex problem for communicating this critical information about a palliative patient experiencing distress and/or challenges with decision-making. Related to a suffering/coping continuum scale, a participant observed:

> ...a religious/spiritual coping score... And I'm conflicted about even putting a score, because I don't think it's quite that simple. But, what I mean by that is the information of what within their spiritual orientation is causing them challenges right now and what within their spiritual orientation is helping them and how does that kind of balance out.

Connected with, but different from a suffering/coping scale, one of the interprofessionals stated concerning a decision-making scale as well:

> You know, you could almost have two scales, one of spiritual distress, like why did God do this to me? Why are my prayers not being answered that way I want them to be? You know, that kind of spiritual, real spiritual distress that's questioning their spirituality or suffering from that. And then there could almost be a different scale of, how important is your spirituality to your medical decision making, from 0–10? Right, because some people are like, not at all important. Some people like, eh, it's involved, and then for some people, it's 10 out of 10 important.

3.2.4 Recommendations

Participants also identified a clear need for help with language, especially with families who may have a more defined religious or cultural expression at variance from a dominant group. Requests varied from the simple hope of knowing religious leader's titles (imam, elder, pastor) within a certain community to assist in understanding and language for identifying, for example, a specific Christian expression of faith from which a mandate of waiting for a miracle may have arisen. The participants amplified how high the stakes can be in religiously unknown territory with emotionally heightened circumstances. A need was identified for language that may help staff navigate critical moments and conversations, rather than potentially sabotage them. Concerning the subtheme of language, a participant stated:

> I'm looking for help with language. We're called to meet families usually in a time pressured situation when [the patient] is on the brink of death which is the worst crisis they've probably ever seen. And so, we, the language is so important. You make a mistake like that, you may not get a chance to recover or they're so stressed they fire you and you're done and you can't help anymore. So, part of what I'm looking for is language.

Additionally named as a subtheme among recommendations was citing important religious or spiritual practices for the staff, namely, other team members and bedside nursing, to help the patient and family on any given day and especially in time of more acute spiritual need. One of the participants illustrated an imagined situation of this subtheme's significance:

> ...if the chaplain has identified a coping skill that patient has, to encourage the staff to say, you know, in my interview with Mr. Smith, they coped well with prayer and so if they're struggling with anxiety at 2 o'clock in the morning, the nurses can say maybe, you know, would it be helpful if someone came to pray with you.

3.2.5 Ongoing Needs/Goals of Care/Action Plan

Focus group participants also expressed a fifth summary content theme as their wish to see some listing of how ongoing spiritual needs are being addressed, what goals of care the chaplain has in relation to those needs, and what is the plan of care to address those needs and meet those goals. One of the participants pointedly described in what way this could be helpful: "...this is why I'm seeing the family, this is what I'm working on, and this is how it resolved."

Another palliative interprofessional team member offered:

> I had plan as well and then maybe along with that I don't know too, what are the ongoing spiritual needs that need to be addressed? Whether that's in the plan or not, but I had that too, talk to me about spiritual needs.

The significance of the palliative chaplain detailing spiritual needs and having related goals of care described in a structured, templated progress note connects back the importance of the specialist role the chaplain plays on a palliative care team related to religious, spiritual, and existential aspects of care concerning a patient's serious illness.

4 Discussion

We examined the views and perceptions of the non-chaplain palliative care team members. The results illustrate that decision-making is the theme where information and insight about a patient's religion/spirituality as well factors concerning a patient's story, family, and perception of emotion are most desired. While it was surprising that this was the theme most discussed by non-chaplain palliative team members, it was anticipated to be in the robust mix of themes (Wirpsa et al. 2019). The literature points to the significance of religion/spirituality for decision-making at the end of life, especially in an intensive care unit (Ernecoff et al. 2015). Decision-making in stressful circumstances is something palliative care teams work to lessen the burden of with a patient and family. There is also a financial cost lessened when patients near death experience high spiritual support. They are three times more likely to enter hospice compared to those with low spiritual support (Balboni et al. 2010). Additionally, those receiving poor spiritual support from a medical team cost on average $2441 (US dollars) more in the final week of life compared to those who are well-supported (Balboni et al. 2011).

Of secondary importance, non-chaplain palliative care team members want a chaplain's progress notes to address patient suffering as well as to know how the patient is coping with it. Like decision-making, there is a desire to know how a patient's suffering and coping is influenced by religion/spirituality, specifically, as well as by a patient's story, family, and perception of emotion. Addressing religious/spiritual suffering is important as it has been found, for instance, in nearly half (44%; 40/91) of an advanced cancer population at MD Anderson Cancer Center. Also, among those with "spiritual pain" (p. 986), they had worse physical symptoms, apart from shortness of breath, as compared to those without spiritual pain (Delgado Guay et al. 2011). In other words, there is some evidence that, by addressing religious/spiritual suffering, embodied suffering can also be reduced. Focus group members also desired to know how coping, overall, and what orienting system (Pargament 2001) patients and families use to assist them in adjusting to disruptive and uncertain circumstances.

An additional surprise to the analysis process was the extent to which non-chaplain team members wanted a sense of a chaplain's perception of emotion in the narrative progress notes related to a patient and family's response to the serious illness. While related to coping, this theme was distinctive as it was identified more to be helpful for a non-chaplain palliative team member's preparation for and possible approach to care with that patient and family. While feelings were described as a part of this sense of emotion, there was also a sense that it was a perception of what was significant (Furtak 2005) for a patient.

The fourth theme worth highlighting, and not previously mentioned in the literature about a palliative chaplain's progress note, is the practical need for an efficient and weighted summary of content at the top. A participant captured this desire with, "…you don't need to scroll down or look for buried treasure." Meriting a focused discussion about what treasure needs to be at the top of the progress note, as it is not apparent in the literature, is the need for a synthesis, scales, and recommendations. Concerning the synthesis, a palliative chaplain's determination of what descriptive content emerged as most important begs for a brief description in this summary section. Related to scales, Stefanie Monod and team (2010, 2012) in Switzerland have developed and validated the clinically based Spiritual Distress Assessment Tool (SDAT) within a geriatric rehabilitation hospital setting. It could be argued that SDAT cannot be generalized to other countries or to a palliative population, but it is the only known comprehensive spiritual distress or suffering tool used by chaplains in collaboration with patients. A consideration could be whether a one question measure such as "Are you at peace?" (Steinhauser et al. 2006) yields the sense of spiritual suffering and coping desired by palliative care team members. There is no known validated decision-making scale utilized by palliative chaplains in the peer-reviewed literature. There is, however, a growing call for chaplains to have a deepened and broader religious competency (Ragsdale 2018), especially when considering patient decision-making and the request for recommendations to palliative team members regarding religious and/or cultural language to use and not use.

Overall, this investigation yielded a good news/bad news narrative for palliative chaplains. The good news is that there was a robust valuing of the chaplaincy role for partnership in palliative care. This valuing of this role is evidenced by the valued investment of time, energy, and insight of the views and perceptions from busy and burdened clinicians to help improve palliative chaplain spiritual assessment progress notes. Their copious affirmation and generous investment in this process is a takeaway. While not having frequency numbers, it was also verbally acknowledged that palliative chaplain progress notes are read. The bad news, however, is that, "sometimes spiritual care has a reputation for not having a helpful note" and that what was being sought for by palliative partners in the note was often missing. Common sense informs that non-chaplain palliative team members will stop looking for treasure in a chaplain's spiritual assessment progress note if it is not there and/or easily found. Perhaps lending strength to the palliative care team model with the chaplain as a core member were the unsolicited comments offered about the chaplain having a "unique perspective" and generating "insights" that would not have arisen without the chaplaincy/patient conversation. Recognizing chaplains make a "deeper spiritual connection" (Lee et al. 2016), palliative chaplains need to continue to build competency toward constructing progress notes that transfer the outcomes of treasured patient/family relationships to information that has durable value for non-chaplain palliative team members.

4.1   Limitations

It is uncertain whether or to what extent the results are transferable beyond inpatient palliative care teams or the upper Midwest of the United States. Another clear limitation of this study is also that this convenience sample lacked gender, religious, as well as racial/ethnic diversity. A larger (geography, multisite) and more diverse purposive sample could lend more strength to the results. The samples of this research project were interprofessionals primarily based on inpatient palliative care teams, and this arguably influenced their contextualized reflections as compared to hospice-based practitioners or palliative providers from an outpatient context. As with any research project, this one also has undeniable bias. Selection bias is arguably present among those who voluntarily gave of their valuable time and insights to this process. It begs the question of whether the perceptions of those who were not there would have varied the results. Potentially also present was the response bias of social desirability due to Paul's role as chaplain on an inpatient palliative care team for 10 years. Paul had prior professional relationships with some of the participants within the various focus groups.

4.2 Implications

With these contributions from non-chaplain palliative interprofessionals, there is a clear recommendation that chaplains continue to hone Domain 5 (religious, spiritual, existential) specialized information in a way that both quickly and thoroughly communicates how these data best inform palliative partners about decision-making and the suffering/coping experience of seriously ill patients and their loved ones.

An expansion of this project could also include the context of outpatient palliative care as well as with those non-chaplain staff members serving within the last chapter of palliative care, hospice. While this research did not intentionally differentiate pediatric palliative care responses within this study, another project could exclusively focus on those staff members. An inpatient nursing sample of focus group participants could be another critical source of information as nurses, traditionally, provide the highest volume of referrals (Galek et al. 2009) to chaplains. Last, to build on this research, focus groups of non-chaplain palliative care team members could gather to converse and seek feedback preferences for three or four different palliative chaplain spiritual assessment pre-formatted templates.

5 Conclusion

Bruce Rumbold, an Australian spiritual care specialist, offered, "Spiritual assessment tools should not be used without adequate exploration of the assumptions made. Assessment processes need to be adequately conceptualised and practically relevant" (Rumbold 2007, S60). This project began with the desire to address assumptions by exploring the perceptions of those who arguably most read palliative chaplain progress notes. With this information, palliative care chaplains can be more equipped to be evermore practically relevant to the shared work of enhancing quality of life and mitigating the suffering of those with serious illness. In the end, this work matters most when it makes an improved quality of life difference for all people with serious illness. When palliative care chaplains prioritize the stories of how existential, spiritual, and religious factors influence decision-making and suffering/coping, they not only help reduce that suffering; they uncover treasure that enriches the contribution of each person on the palliative care team.

Acknowledgments Research is ideally conducted among teams integrating their many contributions. Meriting the most gratitude are the research participants without whom this study would not have been possible. A word of thanks is also necessary to the advisory committee for the provision of their timely insight and sage counsel. A final expression of gratitude is extended to these guides and experts, Dick Krueger, PhD; Mary Anne Casey, PhD; Megan Winkler, APRN, PhD; Susan O'Conner-Von, RN-C, PhD; and Lex Tartaglia, DMin, BCC. The extension of their wisdom, scholarship, and willingness to share it with us made this work better.

Funding   Transforming Chaplaincy, funded by the Templeton Foundation, sponsored this project. Paul Galchutt was a Transforming Chaplaincy Research Fellow. Paul has also been a healthcare chaplain for 15 years, 10 of which were on an inpatient palliative care team at the University of Minnesota Medical Center. Paul's interest in palliative care spiritual assessment progress notes was spurred by the summons to develop one (Galchutt 2013, 2016). Judy Connolly was the comoderator and analysis partner for this research project. She has contributed to the development of interprofessional palliative care teams in conjunction with the Fairview Health Services Palliative Care Leadership Center established in 2003 by the Center to Advance Palliative Care. Judy and Paul have been healthcare chaplain colleagues for 12 years.

References

Association of Professional Chaplains. 2015. *Standards of practice for professional chaplains.* Retrieved June 18, 2018 from www.professionalchaplains.org/Files/professional_standards/standards_of_practice/Standards_of_Practice_for_Professional_Chaplains_102215.pdf.

Balboni, T.A., M.E. Paulk, M.J. Balboni, A.C. Phelps, E.T. Loggers, A.A. Wright, et al. 2010. Provision of spiritual care to patients with advanced cancer: Associations with medical care and quality of life near death. *Journal of Clinical Oncology* 28 (3): 445.

Balboni, T., M. Balboni, M.E. Paulk, A. Phelps, A. Wright, J. Peteet, et al. 2011. Support of cancer patients' spiritual needs and associations with medical care costs at the end of life. *Cancer* 117 (23): 5383–5391.

Cassell, Eric J. 2004. *The nature of suffering and the goals of medicine.* 2nd ed. New York: Oxford University Press.

Choi, P.J., V. Chow, F.A. Curlin, and C.E. Cox. 2019. Intensive care clinicians' views on the role of chaplains. *Journal of Health Care Chaplaincy* 25 (3): 89–98.

Delgado-Guay, M.O., D. Hui, H.A. Parsons, K. Govan, M. De la Cruz, S. Thorney, and E. Bruera. 2011. Spirituality, religiosity, and spiritual pain in advanced cancer patients. *Journal of Pain and Symptom Management* 41 (6): 986–994.

Ernecoff, N.C., F.A. Curlin, P. Buddadhumaruk, and D.B. White. 2015. Health care professionals' responses to religious or spiritual statements by surrogate decision makers during goals-of-care discussions. *JAMA Internal Medicine* 175 (10): 1662–1669.

Ferrell, B.R., M.L. Twaddle, A. Melnick, and D.E. Meier. 2018. National consensus project clinical practice guidelines for quality palliative care guidelines. *Journal of Palliative Medicine* 21 (12): 1684–1689.

Fitchett, George. 2017. *Next steps in spiritual assessment.* [Power Point Slides]. Chicago.

Frank, Arthur W. 2010. *Letting stories breathe: A socio-narratology.* Chicago: University of Chicago Press.

Furtak, Rick A. 2005. *Wisdom in love: Kierkegaard and the ancient quest for emotional integrity.* Notre Dame: University of Notre Dame Press.

Galchutt, Paul. 2013. A palliative care specific spiritual assessment: How this story evolved. *OMEGA-Journal of Death and Dying* 67 (1–2): 79–85.

———. 2016. Chaplaincy scope of practice note: The evolution of a specific palliative care spiritual assessment. *PlainViews* 13: 12.

Galek, K., L.C. Vanderwerker, K.J. Flannelly, G.F. Handzo, J. Kytle, A.M. Ross, and S.L. Fogg. 2009. Topography of referrals to chaplains in the metropolitan chaplaincy study. *Journal of Pastoral Care and Counseling* 63 (1–2): 1–13.

Krueger, Richard A., and Marie A. Casey. 2015. *Focus groups: A practical guide for applied research.* 5th ed. Thousand Oaks: Sage.

Lee, B.M., F.A. Curlin, and P.J. Choi. 2016. Documenting presence: A descriptive study of chaplain notes in the intensive care unit. *Palliative & Supportive Care* 15 (2): 190–196.

Monod, S.M., E. Rochat, C.J. Büla, G. Jobin, E. Martin, and B. Spencer. 2010. The spiritual distress assessment tool: An instrument to assess spiritual distress in hospitalised elderly persons. *BMC Geriatrics* 10 (1): 88.

Monod, S., E. Martin, B. Spencer, E. Rochat, and C. Büla. 2012. Validation of the spiritual distress assessment tool in older hospitalized patients. *BMC Geriatrics* 12 (1): 13.

National Consensus Project Clinical Practice Guidelines for Quality Palliative Care. 2013. *Clinical practice guidelines for quality palliative care.* 3rd ed. Retrieved June 18, 2018 from https://www.nationalcoalitionhpc.org/ncp-guidelines-2013.

Pargament, Kenneth I. 2001. *The psychology of religion and coping: Theory, research, practice.* New York: Guilford Press.

Puchalski, C., B. Ferrell, R. Virani, S. Otis-Green, P. Baird, J. Bull, et al. 2009. Improving the quality of spiritual care as a dimension of palliative care: The report of the Consensus Conference. *Journal of Palliative Medicine* 12 (10): 885–904.

Ragsdale, J.R. 2018. Transforming chaplaincy requires transforming clinical pastoral education. *Journal of Pastoral Care & Counseling* 72 (1): 58–62.

Rumbold, B.D. 2007. A review of spiritual assessment in health care practice. *Medical Journal of Australia* 186: S60–S62.

Steinhauser, K.E., C.I. Voils, E.C. Clipp, H.B. Bosworth, N.A. Christakis, and J.A. Tulsky. 2006. "Are you at peace?": One item to probe spiritual concerns at the end of life. *Archives of Internal Medicine* 166 (1): 101–105.

Tartaglia, A., D. Dodd-McCue, T. Ford, C. Demm, and A. Hassell. 2016. Chaplain documentation and the electronic medical record: A survey of ACPE residency programs. *Journal of Health Care Chaplaincy* 22 (2): 41–53.

Wirpsa, J.M., E.R. Johnson, J. Bieler, L. Boyken, K. Pugliese, E. Rosencrans, and P. Murphy. 2019. Interprofessional models for shared decision making: The role of the health care chaplain. *Journal of Health Care Chaplaincy* 25 (1): 20–44.

Charting Spiritual Care: Ethical Perspectives

Guy Jobin

1 Introduction

The way in which healthcare institutions operate clearly has an influence on the provision of spiritual care. Clinical culture, i.e. all the knowledge, techniques, attitudes and values mobilized to provide care to patients in a healthcare institution or system, has an impact on several aspects of spiritual care, as do the administrative rules that govern each institution. In short, the provision of spiritual care in an institution or healthcare system is shaped by the normative and administrative framework of the prevailing institutional environment. This trend, inherent to the growing integration of spiritual care within the clinical system, illustrates spiritual care's ability to adapt but also its "precariousness." Its fragility requires chaplains to participate in processes that may change and transform their own approach to spiritual support.

Computerization is one of the changes that have forced chaplains to adopt new ways of communicating the information needed by the care team. This is always done with the goal of providing the best possible care, i.e. care that is appropriate and adjusted to the needs and expectations of patients and their families. As we will see, the new ways of sharing information have effects that go beyond a mere functional improvement to the communication of spiritual and religious information within the interdisciplinary care team. The computerization of patient files and charting processes may have unexpected consequences and raises a number of questions about the conditions for providing care and the quality of the spiritual care received by patients and their families.

The specialized literature presents a unanimous view: documenting the particularities of a specific clinical case helps improve care (Tartaglia et al. 2016). This observation holds true especially in the case of electronic health records (EHRs),

G. Jobin, (✉)
Laval University, Quebec City, QC, Canada
e-mail: Guy.Jobin@ftsr.ulaval.ca

S. Peng-Keller, D. Neuhold (eds.), *Charting Spiritual Care*,
https://doi.org/10.1007/978-3-030-47070-8_12

which optimize the consultation of medical records for the purpose of verifying quality of care and conducting research to improve practices (King et al. 2014).

Chaplains often add notes to a patient's medical record, a process known as "charting." The practice established itself quickly in some jurisdictions, and more slowly in others (such as Québec, in particular), but is now part of the clinical landscape. Its importance is recognized by both chaplains' organizations, which see it as a skill to be acquired, developed and perfected (Board of Chaplaincy Certification 2017),[1] and by organizations that accredit and certify care institutions (JCAHO 2019).[2] As charting by chaplains has become an integral part of clinical life, it has also become subject to the general trend among care institutions to dematerialize medical, institutional and personal records.

The few studies that exist on the drafting of EHR notes by chaplains show that they are committed to the practice (Goldstein et al. 2011; Lee et al. 2017; Tartaglia et al. 2016), which is not surprising, given that it contributes to the goal of professionalization for spiritual care providers in the Western world (VandeCreek and Burton 2001; Handzo et al. 2014). The right to add notes to patient records was a demand that chaplains had been making for several decades in pursuit of greater recognition as professionals, in other words as peers within the interdisciplinary care team (Ruff 1996). EHR access consolidated their professional status. The fact that chaplains support the current practice, as noted by researchers, shows that there is little or no resistance to the trend if electronic tools are in place in the care environment.

EHR implementation raises ethical issues that must be addressed. An ethics-based review of information communication technologies must highlight the fact that they involve the same ethical challenges as the practices they replace and must also identify the challenges specifically raised by the use of technological innovations in the clinical environment. This is the goal of this chapter. The discussion is divided into four parts: (1) the ethical issues surrounding the communication of information, (2) the security of system access, (3) the impact of technology on clinical judgment and (4) the issue of recognition for the people involved in the care relationship.

Before examining these ethical issues directly, it is important to discuss some of the elements in the normative and professional context for EHR use by chaplains.

---

[1] professionalchaplains.org/files/2017%20Common%20Qualifications%20and%20 Competencies%20for%20Professional%20Chaplains.pdf, retrieved March 1, 2019. The list of qualifications and skills has been adopted by five Canadian and American chaplaincy associations. Among the skills listed: PPS11 Document one's spiritual care effectively in the appropriate records.

[2] Joint Commission for the Agreement of Healthcare Organisation (JCAHO), www.jointcommission.org/standards_information/jcfaqdetails.aspx?StandardsFaqId=1492&ProgramId=46m, retrieved March 1, 2019.

2   Normative and Professional Context

In the complex world of healthcare institutions, where practically everything is regulated by technoscience and bureaucracy, spiritual care is part of an environment structured by many different legal, administrative, professional, ethical and conduct-related standards. The normative framework may vary from place to place, but some degree of regulation is found everywhere. In fact, the normative environment stems from, and is supported by, two sets of standards. The first comes from "outside" the world of spiritual care and consists of the standards enacted by health certification authorities[3] and the legislation in force in the jurisdiction concerned,[4] along with any other rules created by the legitimate powers and authorities.[5]

The second set of standards originates "inside" the world of spiritual care and includes the standards of practice defined by the professional chaplaincy associations.[6] In addition to these standards, professional associations set standards for training and the skills needed to provide spiritual care. For example, under the *Common Qualifications and Competencies for Professional Chaplains*, a chaplain must possess the professional competency to "Formulate and utilize spiritual assessments, interventions, outcomes, and care plans in order to contribute effectively to the well-being of the person receiving care."[7]

Both sets of standards contribute towards the professionalization of spiritual care. They reflect, in their own way, changes to the task of chaplaincy, which has gone from a vocational calling to a professional commitment. Chaplaincy has adapted to the changes in healthcare that have occurred in recent decades and, more generally, to the social transformation brought about by ethical and religious pluralization in Western countries and the accompanying disaffection with the monotheistic religious traditions and the groups that embody them.

---

[3] As noted above, the JCAHO in the United States, Accreditation Canada, etc.

[4] For example, the Civil Code of Québec; the Health Insurance Portability and Accountability Act (HIPAA) in the United States; the Regulation of the European Parliament and the Council of 27 April 2016 on the protection of natural persons with regard to the processing of personal data and on the free movement of such data (https://eur-lex.europa.eu/legal-content/EN/TXT/PDF/?uri=CELEX:32016R0679&from=FR); the Loi Informatique et libertés (France); etc.

[5] For example, the religious regulation framework adopted by a jurisdiction, charters of rights and freedoms, and any agreements between a hospital administration and a specific religious group.

[6] Such as the Canadian Association of Spiritual Care, Association des Intervenants et Intervenantes en soins spirituels du Québec, Association of Professional Chaplains (USA), etc.

[7] Common Qualifications and Competencies for Professional Chaplains, PPS10. To which must be added the competency of communicating results, noted above, cf. note 2.

3   Ethical Perspectives

As indicated in the introduction, four ethical issues will be examined here: (1) the
ethical issues surrounding the communication of information, (2) the security of
system access, (3) the impact of technology on clinical judgment and (4) the issue
of recognition for the people involved in the care relationship. Readers will note that
the ethical issues involved are not just issues of deontology and have a far broader
scope than the standards that govern the professional task of providing patients with
spiritual support. It is because they directly or indirectly affect the quality of clinical
judgments and the decision-making process that these themes are examined here
in detail.

3.1   Ethical Issues Surrounding the Communication
of Information

The ethical issues surrounding the communication of information in a clinical setting did not emerge suddenly at the same time as EHR. Confidentiality, professional
secrecy, and all aspects of the disclosure or non-disclosure by caregivers of information gathered from patients as part of the professional relationship, are all mentioned in the Hippocratic Corpus (Beauchamp and Childress 2009, 303). However,
the emergence of a clinical culture based on the ideal of interprofessional collaboration significantly changed the rules governing confidentiality. Since the 1980s, the
strict interpretation of medical privilege has not been matched in actual practice,
and the pressure placed by recent changes in clinical culture on the normative
framework for confidentiality that applies to care professionals has a parallel in the
field of chaplaincy.

The fundamental question for professional secrecy arises at the point of intersection between two key normative realities: the patient's right to privacy and the needs
of society, which include the need of the system and practitioners to ensure effective
care delivery (Verdier 2007). This point is, in fact, a place of tension between professional secrecy and the "imperative" need of care professionals to share information (Gilbert and Mettler 2010; Pautier 2017; Gekière and Soudan 2015). At a time
when interprofession collaboration within care teams is being consolidated, the
sharing of private information is key to the efficient and effective coordination of
care, provided the delicate balance between the rights and duties of all players in the
care relationship is respected. Although the tension between privacy and public
interest is well known and discussed in the fields of medical ethics and bioethics, it
becomes even more acute when new technologies multiply the possibilities for
accessing medical information and promote sharing between ever greater numbers
of stakeholders.

The ethical issues surrounding communication can be encapsulated in a few key
questions: What information should a chaplain enter in an EHR? What criterion or

criteria should be used to sort all the information gathered? What justification should be required if a chaplain communicates a piece of information? What should be done with information that could compromise the patient's insurability or influence the quality of the care relationship? To these questions we can add those concerning EHR access, which are addressed in Sect. 3.2.

For chaplains, the principle of charting is already accepted, since it is important to "leave footprints" (Ruff 1996), given that if it isn't charted, it didn't happen, according to the clinical adage. The "professional" imperative to document the spiritual support process arises as soon as the chaplain is required to "report" his or her actions to colleagues in the interdisciplinary care team. Although not needed to fulfil a reporting or accountability requirement, a note added to the file lets other clinical workers know what work the chaplain has done and what recommendations the chaplain has for the team.

An example will complete this examination of the ethical issues surrounding the communication of information using EHR: Ruff's proposal, although dated, nevertheless presents the question of confidentiality effectively from the chaplain's point of view (Ruff 1996, 389–390). Ruff uses principles such as the right to confidentiality and the right to privacy to establish a distinction between information collected under the confessional seal and information gathered during a "normal conversation" that is not *sub secreto*. The strict rules governing confession suffer no exceptions, while information from a conversation can be entered in the record unless the patient objects: "Chaplains must distinguish between information shared by a patient in the normal course of conversation and information shared under the confessional seal. The former, with appropriate pastoral sensitivity, can be charted. The latter should never be charted" (Ruff 1996, 389). If the chaplain has doubts, or if the patient has expressed reservations, the situation must be discussed and the chaplain must always respect the patient's wishes. Ruff's position is clearly in favour of sharing information with team members, but he specifies that "information can always be shared with other staff verbally if the chaplain feels it is appropriate or too sensitive to be included in the chart" (Ruff 1996, 390). The principles and rules presented by Ruff are still present in the latest versions of the codes of conduct for chaplains produced by professional associations such as the Association of Professional Chaplains (USA) and the Canadian Association for Spiritual Care and by associations that supervise training programs, such as the Association for Clinical Pastoral Education. All of these regulatory bodies expect that the "assessment and summary of a chaplain's care is documented" (Common Standards for Professional Chaplaincy 2004). There is therefore a professional duty to add notes to the file, whatever its format – a duty created by the professionalization of the actions that provide spiritual support during illness. The Canadian Association for Spiritual Care (CASC) states, in its 2016 revised *Code of Ethics*, that the spiritual care practitioner "provide[s] other professionals with chart notes where they are used that further the treatment of the clients or patients, obtaining consent when required" and that s/he "communicate[s] sufficient information to other care team members while respecting the privacy of clients."[8]

We can end this section by mentioning that the normative framework for the disclosure of information gathered by chaplains leaves room for professional judgment when determining what must be communicated, to whom, at what time, by what means and with what degree of consent from the patient. The relevant standards give chaplains a responsibility that, as it were, precedes the act of adding a note to the file. This is the responsibility to discern what can be disclosed and to obtain the patient's consent to do so. In short, the responsibility is the same for any information entered in a file, whether paper-based or computerized.

3.2 Security of Computer System Access

This issue extends beyond the question of confidentiality between peers, to encompass institutional protection for the data gathered using the digital platforms made available to care teams. In addition to protection against piracy from outside the institution – obviously an issue that is not limited to healthcare institutions – institutions must guarantee that only authorized members of staff can access electronic files. Patients, of course, must have given consent to access (Manaouil 2009). But once these ethical and legal considerations are brought into play, a decision must be made concerning the extent of access. Who can consult the information in the EHR?

Three categories of "readers" other than care personnel come to mind, each with its own access context. The first category is clinical researchers, who may require access to the electronic files, just as they do to paper-based records. It has already been noted that EHRs are a significant source of research data, especially since consultation is facilitated by the digital format of the information stored. An anonymization process is needed before access to data and records is allowed. However, some jurisdictions prohibit the use of EHRs for epidemiological studies[9] and limit access to purposes connected with the clinical coordination of care.

The second category is clinical training and covers interns in chaplaincy training programs. For these interns, training for their future profession necessarily relies on access to the information in the records, provided they, like other clinicians, undertake to comply with the usual rules on data confidentiality.

The third category poses a more delicate question. It is composed of natural caregivers, volunteers and other non-care personnel who may be closely associated with the care of a sick person, but have no professional qualifications or care mandate from the institution. In general, volunteers do not have access to data, as confirmed by the studies mentioned previously. Besides volunteers, the most difficult

---

[8] CASC, Code of ethics, respectively Section D. #9 and 10; www.spiritualcare.ca/ethics_home/casc-code-ethics, retrieved March 6, 2019.

[9] This is the case in France, cf. Bourdaire-Mignot 2012.

case in this category is that of family members acting as natural caregivers, in other words family members or friends who take responsibility for some of the care provided in the home. The EHR could be a source of relevant information for the care given. There is little documentation on this topic.[10] Chaplains must then be careful about the information they enter in the record when natural caregivers are involved in providing care, especially over a long time period.

3.3   Influence on Clinical Judgments

The use of standardized tools is already widespread in the technoscientific clinical culture and office environment of contemporary healthcare institutions and, as a result, there are calls from the world of chaplaincy itself to allow chaplains to participate in electronic platforms. These calls for closer involvement by chaplains in the clinical process, although they tend to show that healthcare institutions can promote and confirm the objective of professionalization,[11] nevertheless introduce a constraint that can be seen as the price to pay: the growing standardization of charting practices.

Discussion of this ethical issue – because it affects the quality of the communication of the clinical judgment – must begin with the observation that many different researchers, after looking at the question of charting by chaplains, agree with the push for standardization, in other words the adoption of a "charting model" (Peery 2012).[12] This objective of clinical research in the field of spiritual care, known as outcome-oriented chaplaincy, clearly has links to the movement in support of a form of evidence-based chaplaincy, in other words a focus "on outcomes of spiritual care, the development and testing of effectiveness of interventions, the development and evaluation of assessment and screening tools and research about key subgroups of patients" (Damen et al. 2018, 62).

Standardized charting does not automatically lead to a reductive format or, in other words, a reduction in the quality of the information communicated. It may, on the contrary, encourage chaplains to add information to the file that is significant for the decision-making process. This conclusion emerges from the small number of studies already cited: the use of a template acts as a filter for sorting information that is useful from information that is not. Training in charting ensures that chaplains do not add a mass of descriptive detail that may already be recorded elsewhere in the file but, instead, include elements of interpretation and assessment from each spiritual support situation (Lee et al. 2017, 194–195). Adding notes to an electronic file

---

[10] The article by Smeets and de Vries, in this book, indicates that in the Netherlands, patients retain ownership of the data recorded in the file, meaning that the patient and the patient's family have full access to the record. Elsewhere, as in Québec, the record is owned by the healthcare institution.

[11] Recognition is even more significant in jurisdictions where spiritual care is not legally recognized as a profession, such as the Netherlands and Quebec.

[12] Cf. also the contribution by Brent Peery in this book.

offers an opportunity to improve the charting procedure. After all, chaplains' notes may offer a significant contribution to clinical discussion and decision-making (Johnson et al. 2016, 145).

Johnson et al. nevertheless raise a question that is relevant to our discussion. The standardization of charting procedures must not lead to the omission of details that are idiosyncratic by nature (Johnson et al. 2016). They emphasize that charting must allow the inclusion of information that enriches the understanding of the patient's overall situation. Similarly, standardized charting must not be limited simply to an observation of the current state or a generic and even superficial judgment of the case. Each episode recorded must be part of a biographical progression and personal history that is already under way, for which the record is one element of tangible proof. The interpretation of a patient's situation can gain from being re-centred over time on the "long term." This added perspective will certainly improve the accuracy of the decisions made by the clinical team.

In addition, just as the virtues of a narrative approach have been recognized in medical practice (Charon 2006; Charon and Montello 2002), it is clear that a narrative recording of information about the patient for clinical judgment purposes can improve EHR use, provided chaplains are given the opportunity to provide the relevant data. The study by Johnson et al. of free-text documentation by chaplains highlights its impact in improving the decision-making process by recording elements that would be hard to report would otherwise be because they are so deeply rooted in the specifics of the case.

Bringing all the above remarks together, we can suggest that, ideally, any note added to an EHR should provide descriptive elements that are not already included in the record, elements that contribute to the interpretation and assessment of the situation, and elements to situate the clinical case in the patient's specific life history. This is why the choice of the interface model used in the clinical world is crucial, since it will partly determine the quality of the decision-making process and, ultimately, the appropriateness of the decisions made with respect to the actual case. This is confirmed in the study by Rathert et al. about caregiver/patient communications in the era of electronic health records (Rathert et al. 2017, 50–64). In a review article, the authors show that the use of EHR tools produces ambivalent results: "EHR use improves capture and sharing of certain biomedical information. However, it may interfere with collection of psychosocial and emotional information, and therefore may interfere with development of supportive, healing relationships" (Rathert et al. 2017, 62). This means that the standardization of charting practices may have an influence, via the decision-making process, on care practices and outcomes.

3.4 Ethical Issues Surrounding Recognition

The next issue examined here touches on a fundamental question in the field of ethics: recognition of subjectivity, in other words the fact (and normative ideal) of seeing the other as a subject in his or her own right, and not simply as a means or even an object upon which work is performed. A long philosophical tradition, beginning in the late eighteenth century, has addressed the specific question of recognition.[13]

To summarize succinctly, the "problem" of recognition can be defined as "identifying the conditions, processes and situations that lead to a situation in which *Ego* and *Alter* consider each other as equals [in dignity and law] and where the link between them is based on mutual respect and esteem. [...] [The notion of recognition] attempts to encapsulate the origin and consolidation of moral subjectivity in inter-subject relations – within the family, within society – and in the genesis of moral, social and legal relations within society" (Jobin 2013).

Recognition of the other as a subject has an undeniably moral scope since it confirms the other as an autonomous subject. In addition, recognition is expressed in several ways, such as respect for the other's integrity in interpersonal relations and the equality of all in dignity and in law. But beyond the legal aspect, the process of recognition involves everything that defines the particularity and unicity of individuals or groups in a given society and all that "deserves" esteem. Attention paid to the other as an other, as different from one's self, a unique being with a singular life path, is a mark of recognition, and the recognition offered in this way is a mark of mutual respect and esteem between individuals who recognize each other with their own specific differences. The link between the ethical question of recognition and the main focus of this text is based, specifically, on the latter aspect of the theory of recognition, and we can group some of the issues raised in the EHR literature under the theme of the ethics of recognition. The following discussion, without referring directly to chaplains, can shed some light on the topic.

Some criticism has been directed at various potentially problematic aspects of EHR implementation. The issue of anchoring a clinical situation in a life story, which can easily be masked by the standardized information required by the interface, resurfaces in an analysis of the political interests that underlie the implementation of templates. "No doubt, EHR interfaces are designed to make their use as efficient as possible, allowing clinicians to choose between checkboxes and radio buttons for predefined options. As EHRs evolve, these embedded layers add up and increasingly require that clinicians, at the time of care, enter a vast amount of highly structured information, much of which may not be clearly relevant to the individual patient" (Hunt et al. 2017, 406). This criticism by Hunt et al. points to the danger of obscuring the patient's singularity behind the generic information "required" for the EHR. In the end, the authors make the hypothesis that "the EHR enforces the

---

[13] German philosophers including Fichte (1762–1814) and Hegel (1770–1831) addressed recognition specifically. More recently, Jürgen Habermas and, above all, Axel Honneth have returned to the topic.

centrality of market principles in clinical medicine, redefining the clinician's role to be less of a medical expert and more of an administrative bureaucrat, and transforming the patient into a digital entity with standardized conditions, treatments, and goals, without a personal narrative." (Hunt et al. 2017, 418). One unwanted side-effect is the "patient disappearing," just when the goal is to share as much information about the patient as effectively as possible. The irony of the situation is clearly apparent. The patient, as a whole person, is hidden behind the general data generated by the methods used to investigate, gather and share data in an attempt to take responsibility for the patient.[14]

As highlighted by the last quote, the possibility of losing sight of the patient is not just a problem for the patient, but also for clinicians whose role now includes an extra administrative component that may distance them from their primary expertise. Other researchers have made similar observations. Petrakaki et al. claim that the introduction of EPRs (Electronic Personal Records) in England had a clear impact on clinical practice and the cooperative relationship between members of the care team. The research team showed that "EPR affords, as it interacts with healthcare professional practice, some level of standardisation of healthcare professional conduct and practice, curtailment of professional autonomy concerning clinical decision-making and enlargement of nurses' roles and redistribution of clinical work within and across professional boundaries" (Petrakaki et al. 2016, 221). The point that attracts our attention here is the effect on clinical judgment, which can be perceived negatively by some players in the clinical world. Clinicians' autonomy with regard to their clinical judgment and active involvement in the decision-making process for a given patient may be affected by the limits imposed by a computer interface. Recognition for clinicians as subjects in their own right also requires respect for their professional autonomy.

4   Conclusion

The introduction of electronic health records (EHRs) into clinical practice appears to be irreversible. Where EHRs are used, chaplains have cooperated willingly with this way of reporting and sharing information with other members of the care team. They will have to, as a result, adapt their own note-taking practices to ensure effective, relevant and meaningful communication as part of the joint decision-making process.

---

[14] The possibility of the patient disappearing reinforces the need for templates that offer space for entering free text and recording information that will provide a more substantial and contextual view of the patient and his or her situation. The issue of recognition clearly creates tension between effective inter-professional communication, based on formalization and standardization, and an exhaustive approach, which supports appropriate decision-making tailored to the patient's overall situation. I thank Simon Peng-Keller for his pertinent comments on this topic.

Although the specialized literature has addressed some of the "classic" ethical issues raised by EHRs, in particular those in connection with confidentiality and access, other questions, no less crucial, have received less attention and are addressed here. They include questions about the recognition of all players in the care relationship (both patients and caregivers) as subjects and the communication of "non-generic" information about emotions, values, life history, etc. The fact that chaplains contribute to EHRs is both a sign of and a vector for recognition of their work within healthcare institutions – yet a recognition that could involve a price to pay for chaplains and patients.

The classic issues do not acquire a new status because of the features of the computerized tools used – the regulation of professional and deontological conduct is the same as for paper-based tools. However, what requires more attention and discussion is the potential that computerized tools offer for noting the specific features of each situation, which in turn can provide maximally relevant and coherent input for clinical judgment. It is clear that, in everyday clinical practice, the ideal conditions for making a clinical judgment are not always present, but the shortcomings of the tools used to gather and share information should not be allowed to exacerbate to this situation.

Lastly, as with paper records, the fact that chaplains contribute to EHRs is both a sign of and a vehicle for recognition of their work within healthcare institutions. The overall picture that emerges from the literature so far as that the situation must be closely monitored, because recognition could involve an unacceptable price to pay for chaplains and patients. As in many other cases of technological innovation in a clinical setting, the effects on the quality of the clinical relationship – whether a care relationship or a support relationship – must be the main concern for players in the relationship and for the assessment of the practices involved.

References

Association of Professional Chaplains, *Common standards for professional chaplaincy*. 2004. https://www.professionalchaplains.org/Files/professional_standards/common_standards/common_standards_professional_chaplaincy.pdf. Retrieved 3, 2020.

Beauchamp, Tom L., and James F. Childress. 2009. *Principles of biomedical ethics*. 6th ed. Oxford: Oxford University Press.

Board of Chaplaincy Certification Inc. 2017. *Common qualifications and competencies for professional chaplains*. www.professionalchaplains.org/files/2017%20Common%20Qualifications%20and%20Competencies%20for%20Professional%20Chaplains.pdf. Retrieved March 1, 2019.

Bourdaire-Mignot, Camille. 2012. Le dossier médical personnel (DMP): un outil de stockage des données de santé en vue d'un partage. *Revue générale de droit médical*: 295–311.

Charon, Rita. 2006. *Narrative medicine. Honoring the stories of illness*. Oxford: Oxford University Press.

Charon, Rita, and Martha Montello. 2002. *Stories matter. The role of narrative in medical ethics.* New York: Routledge.

Damen, Annelicke, Allison Delaney, and George Fitchett. 2018. Research priorities for healthcare chaplaincy: Views of U.S. chaplains. *Journal of Healthcare Chaplaincy* 24 (2): 62.

Gekière, Claire, and Serge Soudan. 2015. Dossier patient informatisé et confidentialité: évolution des modèles et des pratiques. *L'information psychiatrique* 91 (4): 323–330.

Gilbert, Muriel, and Désirée Mettler. 2010. Confidentialité et partage d'information en soins palliatifs. *Médecine et Hygiène* 25 (3): 105–112.

Goldstein, H. Rafael, Deborah Marin, and Mari Umpierre. 2011. Chaplains and access to medical records. *Journal of Health Care Chaplaincy* 17: 162–168.

Handzo, George, Mark Cobb, Cheryl Holmes, Ewan Kelly, and Shane Sinclair. 2014. Outcomes for professional health care chaplaincy: An international call to action. *Journal of Health Care Chaplaincy* 20 (2): 43–53.

Hunt, Linda M., Hannah S. Bell, Allison M. Baker, and Heather A. Howard. 2017. Electronic health records and the disappearing patient. *Medical Anthropology Quarterly* 31 (3): 403–421.

Jobin, Guy. 2013. Reconnaissance. In *Dictionnaire encyclopédique d'éthique chrétienne*, ed. L. Lemoine, E. Gaziaux, and D. Müller, 1713–1714. Paris: Cerf.

Johnson, Rebecca, M. Jeanne Wirspa, Lara Boyken, Matthew Sakumoto, George Handzo, Abel Kho, and Linda Emanuel. 2016. Communication chaplains' care: Narrative documentation in a neuroscience-spine intensive care unit. *Journal of Health Care Chaplaincy* 22: 133–150.

Joint Commission for the Agreement of Healthcare Organisation (JCAHO). www.jointcommission.org/standards_information/jcfaqdetails.aspx?StandardsFaqId=1492&ProgramId=46m. Retrieved March 1, 2019.

King, Jennifer, Vaishali Patel, Eric W. Jamoom, and Michael F. Furukawa. 2014. Clinical benefits of electronic health record use: National findings. *Health Services Research* 49 (1): 392–404.

Lee, Brittany M., Farr A. Curlin, and Philipp J. Choi. 2017. Documenting presence: A descriptive study of chaplain notes in the intensive care unit. *Palliative and Supportive Care* 15: 190–196.

Manaouil, Cécile. 2009. Le dossier médical personnel (DMP): "autopsie" d'un projet ambitieux? *Médecine et droit* 94: 24–41.

Pautier, Silvère. 2017. Le secret professionnel soignant: un enjeu de démocratie sanitaire entre immanence et aliénation. *Recherche en Soins Infirmiers* 130: 53–67.

Peery, Brent. 2012. Outcome oriented chaplaincy. Intentional caring. In *Professional spiritual and pastoral care: A practical clergy and chaplain's book*, ed. S.R. Roberts, 342–361. Woodstock: Skylight Paths Publishing.

Petrakaki, Dimitra, Ela Klecun, and Tony Cornford. 2016. Changes in healthcare professional work afforded by technology: The introduction of a national electronic patient record in an English hospital. *Organization* 23 (2): 221.

Rathert, Cheryl, Jessica N. Mittler, Sudeep Banerjee, and Jennifer McDaniel. 2017. Patient-centered communication in the era of electronic health records: What does the evidence say? *Patient Education and Counseling* 100: 50–64.

Ruff, Robert A. 1996. "Leaving footprints": The practice and benefits of hospital Chaplains documenting pastoral care activity in patients' medical records. *The Journal of Pastoral Care* 50 (4): 383–391.

Tartaglia, Alexander, Diane Dodd-McCue, Timothy Ford, Charles Demm, and Alma Hassell. 2016. Chaplain documentation and the electronic medical record: A survey of ACPE residency programs. *Journal of Health Care Chaplaincy* 22: 41–53.

VandeCreek, Larry, and Laurel Burton. 2001. Professional chaplaincy: Its role and importance in healthcare. *Journal of Pastoral Care* 55 (1): 81–97.

Verdier, Pierre. 2007. Secret professionnel et partage des informations. *Journal du droit des jeunes* 269 (9): 8–21.

Charting Spiritual Care in Digital Health: Analyses and Perspectives

Simon Peng-Keller

In this final contribution, the main threads of the present volume will be merged and will be brought together, and the prospects for practice and research will be examined. In a first step, I take up the historical approach outlined in the introduction. In order to deepen our understanding of current developments, I identify and describe the main driver behind it. Then, in a kind of *relecture* of the debate, I recapitulate some of the critical and controversial points, and I outline what I take to be the main areas of convergence and the undisputed insights of the contributions to this collection. The paper concludes with desiderata for new research and an outline of the future for the charting of spiritual care.

1 The Genesis and Drivers of a New Chaplaincy Practice

As pointed out in the introduction, charting healthcare chaplaincy in medical records is anything but new. As a professional practice, it had already been developed in the first part of the twentieth century. However, until the digital revolution, it remained a marginal phenomenon. With the rise of EMRs, the framework has changed completely and has opened up new possibilities and challenges for healthcare chaplaincy. Much more than a mere technical adjustment, the introduction of EMRs has had a considerable impact on spiritual care. The changes studied in this book are part of a much larger and ongoing process with as yet unknown consequences: namely, the digitalization of society and healthcare.

Nevertheless, it would precipitate to attribute the new charting of spiritual care solely to digitalization. As the contributions of this volume prove, other drivers have also facilitated the new methods of record keeping. At least three of them are clearly

S. Peng-Keller (✉)
University of Zurich, Zurich, Switzerland
e-mail: simon.peng-keller@uzh.ch

© The Author(s) 2020
S. Peng-Keller, D. Neuhold (eds.), *Charting Spiritual Care*,
https://doi.org/10.1007/978-3-030-47070-8_13

213

identifiable: the emergence of a new paradigm of healthcare chaplaincy; the development of interprofessional spiritual care; and remarkable changes in Western societies concerning the role of religion and spirituality in public spaces. For a more detailed understanding of the evolution examined here, it is worth taking a closer look at these three drivers as well.

(1) For many decades, clinical pastoral education has developed under the influence of the approach developed by Carl Rogers in the early 1940s. Rogers' nondirective counseling has been severely criticized but has until now proved to be helpful. However, it is now being challenged by a new paradigm: outcome-oriented chaplaincy (Hall et al. 2016). Its programmatic claim is reminiscent of a similar turn in psychotherapy, which was triggered by the emergence of effectiveness research and accompanied by heated debate of Klaus Grawe's insistence that psychotherapy must progress from "confession to profession" (Grawe et al. 1994) that resonates perfectly with the goals of outcome-oriented chaplaincy. However, the development of this approach was triggered mainly by financial pressure. With new public management and the ongoing economization of healthcare, cost-benefit considerations reached chaplaincy in the 1990s. At the Barnes-Jewish Hospital in St. Louis, this pressure led Arthur Lucas and his colleagues to develop a new concept of healthcare chaplaincy. The discovery that compassionate presence and purposefulness are not mutually exclusive was a key insight in the learning process described: "In our explorations, we found being present with patients can include informed intentionality." (Lucas 2001, 4) The central argument for outcome-oriented chaplaincy is the inevitability of objectives and critical evaluation. Nondirective presence is appropriate only if it fits with the needs and wishes of patients; and so the patients' needs and wishes must be assessed. After the chaplain's visit, it is then necessary to check if it was helpful. Documenting is necessary for accountability and critical self-evaluation.

(2) While the development of outcome-oriented chaplaincy responds to external factors, especially to economic pressure, interprofessional spiritual care has its roots in twentieth-century reformist healthcare movements as well as in research on spirituality and health (Peng-Keller 2019a). Understood as a common task of all caregivers and volunteers, spiritual care has been, for instance, an essential part of the modern hospice movement since its beginnings in the 1960s. But the issue has been present in other areas of modern healthcare as well. In a groundbreaking resolution in 1984, the WHO recommended that the spiritual dimension should be included in all areas of healthcare. Even though this resolution was accepted only hesitantly, it marked a milestone in the development of interprofessional spiritual care which would begin to institutionalize itself in the 1990s with the establishment of new research positions, training programs, and scientific journals. In recent decades, the empirical literature on spirituality and religiosity has covered more and more areas of healthcare. During the same period, there has also been a gradual integration of spiritual care into curricula and healthcare practice. For example, according to the new learning objectives

for medicine in Switzerland, students must learn to take the spiritual dimension into account within a psychosocial history. When doctors and other health professionals start to assess the spiritual concerns of their patients, the question inevitably arises how these concerns should be noted in the EMR.

(3) The third driver is no less complex, encompassing cultural and mental changes as well as the circulation of religions and spiritualities through globalization and migration. Roughly speaking, it encompasses the spiritual pluralization of Western societies, including a repositioning of organized religion in public life (Peng-Keller 2019b). In (post-)secular and ideologically heterogeneous societies, the status of faith-based chaplaincy in public institutions is controversial. Remarkably, the number of chaplains in the public sector as a whole has grown in recent years, although the public funding of the chaplaincy is contested in many places. Winnifred Fallers Sullivan, in a fine-tuned analysis of this development, has described the new type of chaplain as a "strangely necessary figure, religiously and legally speaking, in negotiating the public life of religion today," as someone who "operates at the intersection of the sacred and the secular, a broker responsible for ministering to the wandering souls of a globalized economy and a public harrowed by a politics of fear" (Sullivan 2014, x–xi).

It is no wonder that in the military and the prison system, the impact of the "new governance in religious affairs" has been particularly significant, as the public interest here is most urgent. In the healthcare sector, the interests are more mixed, and the developments more heterogeneous. Nevertheless, they are to be found in all of the countries and contexts studied above. Particularly telling is the case of Québec, which shows how healthcare chaplaincy has been reshaped in recent decades by changes in society – with consequences for clinical documentation. As we learn from the paper of Bélanger and colleagues, healthcare in Québec was largely Catholic until well into the twentieth century. This slowly began to change in the 1960s. For chaplains, changes within the church could be seen in their new professional title. The traditional designation in Québec "aumônier" was replaced in the 1970s by "agents de pastoral," not least because non-ordained theologians and religious had replaced priests as chaplains. In the 1990s, the term "intervenant(e)s en soins spirituels" was coined. It signified the fact that chaplaincy is not limited to pastoral ministry and religious care. This change was embedded in the reorientation of Québec's politics toward what would be called "open secularism." A report commissioned by the Québec government and written by the sociologist Gérard Bouchard and the philosopher Charles Taylor explains this concept as follows: "Open secularism recognizes the need for a neutral state – legislation and public institutions must not favor a religious or secular view. But it also recognizes the importance that many people attach to the spiritual dimension of life" (Bouchard and Taylor 2008, 140f.). With regard to healthcare, the report calls for a holistic approach that takes into account a person's biopsychosocial and spiritual dimensions. Since a decree issued by the Ministry of Health in 2010, the responsibility for the provision and funding of healthcare chaplaincy lain with the health institutions themselves, which were given the legal obligation to attend to patients' "spiritual needs." Almost exactly the same story, but in a protestant version, could be told about healthcare chaplaincy in the Swiss canton of Vaud. Here as in Québec, the new regulation of the relationship between the state and the churches led to a repositioning of healthcare chaplaincy, a change which also manifested itself in an obligation to keep records.

In sum, the drivers of the rapid development of charting spiritual care in EMRs were manifold. Their complex interactions explain not only its speed but also the heterogeneity of current models and the tensions between the distinct goals of charting chaplaincy work in EMRs. For understandable reasons, these tensions are played down by those who advocate the implementation of this practice. This is one reason to explore them more closely in the following section.

2   Convergences and Controversies

In current discussions about the *who, what, how*, and *why* of recording spiritual care in EMRs, there is a growing convergence on at least four points. First, used as a tool for planning, coordination, and critical self-evaluation, appropriate forms of digital charting can benefit the work of chaplains. Second, it can also have undesired side effects. Third, any future healthcare chaplaincy will have to be a part of the evolving process of digital recording. In a highly specialized professional world where communication and recording is increasingly formed by digital tools, being absent from these new spaces would marginalize the work of chaplains even more. It would mean assimilating the role of chaplains to that of visiting ministers for community members. At least for those who are committed to the project of the clinical pastoral education, there is no alternative to the ongoing process of professionalization which includes standards for recording. Fourth, the ongoing change in healthcare and society, outlined above, forces the chaplaincy to become clearer about its nature and role.

At this point, the controversies begin. Some of them are not new but were already part of the CPE of the twentieth century. Focusing on the *who* of recording, the question of professional identity arises. The sensitive issue here is the proximity of chaplaincy work to psychological counseling and its language. If professional autonomy is to be preserved, distinct profiles are needed. But how can this be guaranteed if the core dimension of spiritual care characterized by particular spiritual traditions and affiliations recedes into the background? The paradigm of outcome-oriented chaplaincy may improve the self-reflectivity and the efficiency of spiritual caregivers, but it can't solve the underlying problem of professional identity, which is exacerbated by the current spiritual-religious diversification. The preceding remarks also seem to touch on matters of identity – and not just practical matters. Who and what are healthcare chaplains in the community of caregivers and in the context of interprofessional spiritual care? If they are themselves healthcare professionals, are they representatives of specific faith communities and spiritual traditions as well? Should they be active members of particular faith communities or rather active members in non-denominational professional associations (or both)? Is it necessary for their work that they are themselves believers and religious practitioners? How important is theological education for the identity of healthcare chaplains? Why not train psychologists in spiritual care so that they can fulfill the role of specialized spiritual carer?

The current development is quite paradoxical: In order to achieve a better professional status in healthcare, one clearly distinguishable from that of the psychologist, chaplains must avoid being exclusively health professionals. What gives chaplains a distinctive profile lies beyond secular healthcare. Thus, they are creatures of two worlds: the world of healthcare, with its rules and quality measures, and the world of a specific spiritual community, with its own language and practices. The first affiliation legitimizes the access chaplains have to the EMR, but it is the second affiliation that affords them their identity. How does this dual affiliation relate to the challenge of charting spiritual care in EMRs? As mentioned in the introduction to this volume, Richard Cabot and Russell Dicks have shown how professional and spiritual aspects are closely intertwined in pastoral documentation. On one level, charting spiritual care is a professional practice with strict rules and determining contexts. However, on a deeper level, the same activity can be understood and cultivated as a spiritual exercise. Cabot and Dicks particularly emphasize the creative side of writing: how the process of documenting can not only clarify an issue but inspire new ideas for future practice.

Thus, the questions of the *what* and the *how* of charting spiritual care are closely connected with the issue of professional identity. If "code switching" is essential for today's healthcare chaplains, as Wendy Cadge (2012) claims, then the practical knowledge of charting is entangled with the orientational knowledge stemming from the perspective of a spiritual view on life and death. As spiritual identities are formed by the language learned in one tradition or another, healthcare chaplaincy has to nourish itself with these languages while navigating with digital skills and codes. For clinical usage in secular and multi-faith contexts, standardized records are indispensable. Preformulated items broad enough to encompass a plurality of spiritual/religious beliefs, practices, and experiences might be a good compromise. However, the flipside of openness is vagueness, something contradictory to the goal of integrating orientational knowledge, the spiritual reframing of the situation of illness, into the charts. So, this task is perhaps best achieved in a free-text section. Narrative entries would serve this purpose best. Chaplains are "bearers of stories" (Anne Vandenhoeck), and theology is accustomed to raising its issues through narrative. But doesn't that also hold true for psychology?

The task of charting spiritual care into EMRs might be seen as a merely technical duty. However, with its questions of the *who*, the *what*, and the *how*, it touches the heart of chaplaincy as a spiritual profession in healthcare. If chaplains have their own way of being "bearers of stories" and of letting "stories breathe" (Frank 2010), then this is because of the shared stories that form their identities. They welcome the stories of the patients in a distinctive horizon, a specific web of significance. The narrow free-text spaces in EMRs might not be the best place for vibrant stories received and empowered within such horizons and webs. But they might at least offer opportunities for hints toward complementary sources of meaning, healing, and support.

3 Perspectives for Further Research

The present volume documents the first two decades of charting spiritual care in EMRs as well as the first decade of research on it. On both levels we are concerned with pioneering work. The experimental and tentative character typical of emerging practices is often reflected in the corresponding research. Being involved in pioneering work is exciting but also challenging. It means navigating a changing field with few reliable landmarks. In this situation, clear strategies for research are all the more urgent. The available literature together with the papers collected here suggests six pathways:

3.1 Empirical Evidence of the Impact of Charting

The first priority for future research should be to investigate the impact of chaplaincy records on caregiving. Given the amount of time and energy it takes to develop and implement new tools and practices, it is essential to prove their appropriateness and usability. On the most general level, this concerns the tools themselves: Are they suitable for the declared aims? Are the goals named attained? And what unintended side effects are there? On a more specific level, the degree and way of implementation of a tool might be examined. If a practice doesn't reach its goals, the problem might lie with the technical tools. But it might be a lack of implementation, training, or supervision. Finally, one might focus on the impact of digital charting on interprofessional communication and contact with patients. Are the chaplaincy entries actually read? Are they understood and perceived as helpful? Do they make a difference to work at the bedside? Empirical evidence on the impact of charting spiritual care may also increase awareness of the potential of a digital practice as well as its limitations.

3.2 Clarifying the Conceptual Framework

This volume has also touched upon research into the conceptual framework of healthcare chaplaincy. As has been shown, the choice of tools and practices of charting always reflects a particular understanding of chaplaincy and spiritual care. Questions around divergent concepts of healthcare chaplaincy lie at a deeper level and call for independent discussion. But, as the development of digital charting tools and practices is shaped by these concepts, it is also necessary to address such questions in this context. Moreover, the effort to clarify the rationale for the participation of chaplains in medical record keeping may shed light on the broader discussion of interprofessional spiritual care. Or, as Bélanger and his colleagues put it, the practice of writing notes in EMR may force us to rethink the theological framework of healthcare chaplaincy. The conceptual decisions, especially, have a legal impact. What legitimizes (or even urges) the chaplaincy to participate in medical recording? If the claim to legitimacy is based on the assumption that the chaplaincy participates in medical care and shares its goals (e.g., pain relief), this has to be justified theologically as well.

3.3   Hermeneutic Analysis of Professional Documentation

The development of digital charting in chaplaincy has brought a neglected topic to the fore: the art of professional documentation in chaplaincy. From a research perspective, hermeneutical, narratological, and technical questions are at stake. Generally speaking, notes in EMR are a specific form of testimony. By means of a highly interpretative process, experiences are transformed into standardized text. This process could and should be studied with the aid of concrete examples. However, the question must also be discussed at a fundamental level: If the professional identity of chaplaincy lies, as is often claimed, in a specific perspective, namely, in the theological view of human life, then this perspective should be recognizable in the digital records of chaplains and in the categories of the tool used. What are the hallmarks of a theological hermeneutic for "living human documents" that can guide the practice of digital charting?

3.4   Ethical Investigations

Until now, ethical investigations concerning the charting of spiritual care have focused mainly on the issue of confidentiality. The discussion of this issue must take account of the different distinctions that are drawn in the various denominational traditions. Catholic theology, for instance, differentiates between the canonically regulated confessional secret and pastoral confidentiality, which are less clearly defined. Both in turn differ from legally regulated professional secrecy. The question of professional identity returns at this stage: As a pastoral profession, chaplaincy is regulated by the concepts and rules of a specific spiritual tradition; as profession in healthcare, the ethical standards and obligations of healthcare professionals are to be applied. Clarifications of these distinctions and overlaps may touch on the deeper question of the relevance and limits of confidentiality in interprofessional spiritual care. Furthermore, as Guy Jobin's contribution to this volume indicates, the charting of spiritual care calls for ethical reflection with respect to a number of different issues and not just confidentiality. Apart from the ubiquitous problem of data security, at least four further questions are to be addressed: What is the impact of technology on care and clinical judgment? How does charting affect the care relationship? Who is accountable for writing and reading records in interprofessional spiritual care? And finally, (how) is the aim of a common culture of values or a cross-professional ethos attainable?

3.5   Charting Spiritual Care in the Context of Digital Health

The question of the inherent dynamics of digitalization in healthcare forms a further research area. Just as digital tools transform the daily practice of physicians and nurses, so they also have profound implications for spiritual care, influencing how chaplains relate to patients, colleagues, and other caregivers. Digital tools shape the organization of work, the forms of communication, the patient's expectations, as well as the infrastructure and the atmosphere of healthcare institutions. As digitalization amounts to a profound transformation of the professional identity of healthcare professionals and chaplains, the changes it brings about must be studied in detail. How will the chaplain's offline ministry be changed and extended by new online spaces for (recording) spiritual care? At a more technical level, there is also the need for more intensive cooperation between IT researchers and practitioners in order to improve the digital tools and their usage.

3.6   Digital Records as a New Tool of Chaplaincy Research

Finally, recording spiritual care in EMRs opens new possibilities for research on spiritual care. As Paul Galchutt and Judy Connolly demonstrate in their contribution, analyzing the digital notes of chaplains can be not only very instructive but also participatory. Apart from the possibility of checking the adequacy as well as the impact of a certain kind of charting, the analysis of digital records with the methods of the digital humanities could provide precious information on the needs of patients, the work of chaplains, and the functioning of interprofessional collaboration. In view of the manifold possibilities which digitized data already offers for studies of all kinds, it is not unlikely that EMRs will be the central instrument for spiritual care research in the future.

4   Outlining the Future

Powered by the three drivers named at the beginning, healthcare chaplaincy is currently surfing through a "sea change" (Massey 2015). The new practice of recording spiritual care in EMRs reflects this change and is itself a catalyst of it. The medium shapes the message as well as the identities of those who use it. This is true for professionals, patients, and relatives. Digital communication between healthcare professionals and patients will intensify in the coming years. Just as telemedicine and cross-clinic EMRs will be interwoven in the future, so may chaplaincy documentation be increasingly intertwined with forms of telechaplaincy (as it has been introduced, for instance, by the influential US Department of Veterans Affairs). "Synthetic situations," in which online and offline spaces overlap (Knorr Cetina 2009), will be increasingly common for healthcare chaplaincy. While it is unlikely that telechaplaincy will replace the direct encounter in the sickroom, it is conceivable that in the future contact will be established digitally. It goes without saying that telechaplaincy, if part of the therapeutic concept, will also have to be recorded in EMRs.

Considering the current trends, the inclusion of interprofessional spiritual care in medical care is likely to become more widely established in the future. To the extent that chaplains participate in the medical treatment mandate, they are then to be regarded as health professionals and subjected to the same obligations. However, this is not a conclusive answer to the question of their professional identity as specialized *spiritual* caregivers. If spirituality has to do with transcendence and ultimate values, and if these are cultivated by distinct traditions and communities, then the identity of chaplains is a matter of their spiritual aim, of the ability to help patients navigate liminal spaces: in the realm of between life and death, between a meaningful cosmos and its collapse, and not least in the border zones between established faith communities and the individual's search for meaning. For these reasons, the "spiritual concerns" of patients and relatives must be afforded space in the online reality of EMRs.

References

Bouchard, Gérard, and Charles Taylor. 2008. *Fonder l'avenir. Le temps de la conciliation.* Rapport de la Commission de consultation sur les pratiques d'accommodement reliées aux différences culturelles. Québec. www.mce.gouv.qc.ca/publications/CCPARDC/rapport-final-abrege-fr.pdf.

Cadge, Wendy. 2012. *Paging god. Religion in the Halls of medicine.* Chicago: University of Chicago Press.

Frank, Arthur W. 2010. *Letting stories breathe. A socio-narratology.* Chicago: University of Chicago Press.

Grawe, Klaus, Ruth Donati, and Friederike Bernauer. 1994. *Psychotherapie im Wandel. Von der Konfession zur Profession.* Göttingen: Hogrefe.

Hall, Eric J., George H. Handzo, and Kevin Massey. 2016. *Time to move forward. A new model of spiritual care to enhance the delivery of outcomes and value in health care settings,* New York. https://spiritualcareassociation.org/docs/resources/time_to_move_forward_report_2016-06-07.pdf.

Knorr Cetina, Karin. 2009. The synthetic situation: Interactionism for a global world. *Symbolic Interaction* 32 (1): 61–87.

Lucas, Arthur. 2001. Introduction to *The discipline for pastoral care giving. Journal of Health Care Chaplaincy* 10: 1–33.

Massey, Kevin. 2015. Surfing through a sea change: The coming transformation of chaplaincy training. *Practice: Formation and Supervision in Ministry* 35: 144–152.

Peng-Keller, Simon. 2019a. Spiritual Care im Gesundheitswesen des 20. Jahrhunderts. Von der sozialen Medizin zur WHO-Diskussion um die ‹spirituelle Dimension›. In *Spiritual Care im globalisierten Gesundheitswesen. Historische Hintergründe und aktuelle Entwicklungen*, ed. Simon Peng-Keller and David Neuhold, 13–71. Darmstadt: Wissenschaftliche Buchgesellschaft Darmstadt.

———. 2019b. Genealogies of 'spirituality'. An historical analysis of a travelling term. *Journal for the Study of Spirituality* 9 (2): 86–98.

Sullivan, Winnifred F. 2014. *A ministry of presence. Chaplaincy, spiritual care, and the law.* Chicago: University of Chicago Press.

Index